Studies in Systems, Decision and Control

Volume 17

Series editor

Janusz Kacprzyk, Polish Academy of Sciences, Warsaw, Poland
e-mail: kacprzyk@ibspan.waw.pl

About this Series

The series "Studies in Systems, Decision and Control" (SSDC) covers both new developments and advances, as well as the state of the art, in the various areas of broadly perceived systems, decision making and control- quickly, up to date and with a high quality. The intent is to cover the theory, applications, and perspectives on the state of the art and future developments relevant to systems, decision making, control, complex processes and related areas, as embedded in the fields of engineering, computer science, physics, economics, social and life sciences, as well as the paradigms and methodologies behind them. The series contains monographs, textbooks, lecture notes and edited volumes in systems, decision making and control spanning the areas of Cyber-Physical Systems, Autonomous Systems, Sensor Networks, Control Systems, Energy Systems, Automotive Systems, Biological Systems, Vehicular Networking and Connected Vehicles, Aerospace Systems, Automation, Manufacturing, Smart Grids, Nonlinear Systems, Power Systems, Robotics, Social Systems, Economic Systems and other. Of particular value to both the contributors and the readership are the short publication timeframe and the world-wide distribution and exposure which enable both a wide and rapid dissemination of research output.

More information about this series at http://www.springer.com/series/13304

Leonid Shaikhet

Optimal Control of Stochastic Difference Volterra Equations

An Introduction

Springer

Leonid Shaikhet
Department of Higher Mathematics
Donetsk State University of Management
Donetsk
Ukraine

ISSN 2198-4182 ISSN 2198-4190 (electronic)
Studies in Systems, Decision and Control
ISBN 978-3-319-13238-9 ISBN 978-3-319-13239-6 (eBook)
DOI 10.1007/978-3-319-13239-6

Library of Congress Control Number: 2014955322

Mathematical Subject Code Classification (2010): 37C75, 93D05, 93D20, 93E15

Springer Cham Heidelberg New York Dordrecht London

Printed on acid-free paper

Springer International Publishing AG Switzerland is part of Springer Science+Business Media (www.springer.com)

Preface

The aim of this book is an introduction to the mathematical theory of optimal control for a relatively new class of equations: stochastic difference Volterra equations of neutral type. Equations of such type arise as independent objects of research, as mathematical models of systems with discrete time, or as difference analogues of stochastic differential and integral Volterra equations of neutral type [9, 109] by their numerical simulation.

These equations belong to the more general class of hereditary systems (also called systems with aftereffect, or systems with memory, or equations with deviating arguments, or equations with delays, or equations with time lag, and so on) and describe the processes whose behavior depends not only on their present state but also on their past history [9, 40, 42, 65–69, 72, 96, 103–106, 109, 138, 153, 154, 178, 179, 187, 199]. Systems of such type are very popular in researches and are widely used to model processes in physics, mechanics, automatic regulation, economy, finance, biology, ecology, sociology, medicine, etc. (see, e.g., [11, 15, 19, 21–23, 26, 27, 33, 34, 53, 62, 75, 86, 115, 126, 132, 144, 148, 152, 162, 166–169, 177–181, 188, 192, 197, 198]).

The general theory of difference equations, including difference equations with delay is represented in [2, 3, 19, 25, 38, 39, 56, 59, 95, 98, 121, 123, 135, 151, 156, 178, 183]. A lot of works are devoted, in particular, to investigation of integral and difference Volterra equations [4, 5, 10, 17, 18, 32, 41, 49–52, 54, 55, 57, 58, 60, 64, 76, 81, 93, 94, 101, 102, 108, 117, 118, 133, 143, 145, 157, 189, 190, 193, 213], to numerical analysis of systems with continuous time by virtue of appropriate difference analogues [1, 35–39, 73, 83, 123, 129, 136, 137, 194, 195, 199], to description of different mathematical models of systems with discrete time [6, 7, 44–46, 78, 79, 82, 91, 92, 127, 128, 131, 139, 149, 150, 155, 206–209].

A very important and popular (in particular, in the last years) direction in the theory and different applications of systems of such type is the optimal control theory. To problems of existence and construction of optimal control in the sense of the given performance criterion and stabilization problems for deterministic and stochastic systems with continuous or discrete time, both without and with delays, plenty of works are devoted [8, 9, 12–14, 16, 20, 23, 24, 28–30, 32, 43, 47, 48, 61, 63, 68, 74, 77–80, 82, 84, 85, 87–91, 99, 109, 112–114, 116, 119, 122, 124, 125, 130, 134, 140–142, 146, 147, 149, 150, 155, 158–161, 163–165, 170–176, 185, 186, 191, 196, 200, 201, 203–212]. There are also many works devoted to problems of optimal estimation (see, for instance, [17, 18, 31, 97, 100, 107, 111, 120, 182, 202]).

In this book, consisting of six chapters, all the above-mentioned basic problems of the mathematical theory of optimal control and optimal estimation are extended on stochastic difference Volterra equations of neutral type. It is shown, in particular, that the difference analogues of the solutions of optimal control problems and optimal estimation problems obtained for stochastic integral Volterra equations cannot be optimal solutions of corresponding problems for stochastic difference Volterra equations.

The introductory Chap. 1 presents an origin of stochastic difference Volterra equations of neutral type, a necessary condition for the optimality of a control for abstract optimal control problem and some auxiliary definitions and lemmas.

In Chap. 2 a necessary condition for the control optimality of nonlinear stochastic difference Volterra equation is obtained and via this condition a synthesis of the optimal control for a linear-quadratic problem is constructed. A synthesis of the optimal control means the control constructed as a feedback control. Some demonstrative examples of calculating the optimal control in a final form are shown.

In Chap. 3 the problem of construction of successive approximations to the optimal control of the stochastic quasilinear difference Volterra equations with quadratic performance functional is considered. An algorithm for constructing such approximations is described. It is shown that successive approximations can be considered both as a program control and as a feedback control.

In Chap. 4 the problem of the optimal stabilization for a linear stochastic difference Volterra equation and quadratic performance functional is considered. The optimal control in the sense of a given quadratic performance functional that stabilizes the solution of the considered equation to mean square stable and mean square summable is constructed. For the quasilinear stochastic difference Volterra equation with quadratic performance functional a zeroth approximation to the optimal control is constructed that stabilizes the solution of the considered difference equation to mean square stable and mean square summable.

In Chap. 5 the filtering problem is formulated. More exactly, the problem of constructing the optimal (in the mean square sense) estimate of an arbitrary partially observable Gaussian stochastic process from its observations with delay is considered. It is proved that the desired estimate is defined by a unique solution of the fundamental filtering equation of the Wiener–Hopf type. Qualitative properties of this equation are discussed and several cases where it can be solved analytically are considered. The relationship between the observation error and the magnitude of delay in observations is investigated. It is shown that the fundamental filtering equation describes also the solutions of the forecasting and interpolation problems. In the case where the unobservable process is given by a stochastic difference Volterra equation an analogue of the Kalman–Bucy filter is constructed: the system of four stochastic difference equations defined the optimal in the mean square sense estimate.

In Chap. 6 two different methods for solution of the optimal control problem for partly observable linear stochastic process with a quadratic performance functional are proposed: the separation method and the method of integral representations. An ε-optimal control of the optimal control problem for a quasilinear stochastic difference Volterra equation and a quadratic performance functional is constructed. A special method is proposed for solution of the optimal control problem for stochastic linear difference equation with unknown parameter and quadratic performance functional. The optimal control in final form is obtained. Numerical calculations illustrate and continue the theoretical investigations.

The bibliography at the end of the book does not pretend to be complete and includes some of the author's publications and publications of his coauthors [31–34, 99, 100, 107–109, 116–120, 171–176, 182] as well as the literature used by the author during his preparation of this book.

The book is mostly based on the author's results and ideas, closely related with the results of other researchers in the theory of difference equations, the theory of hereditary systems, and the optimal control theory [2, 14, 19, 47, 55, 59, 68, 79, 103–106, 112, 130, 183]. It is addressed to experts in the mathematical optimal control theory as well as to a wider audience of professionals and students in pure and applied mathematics.

Taking into account that the possibilities for further improvement and development are endless the author will appreciate receiving useful remarks, comments, and suggestions.

Donetsk, Ukraine Leonid Shaikhet

Contents

Chapter 1
Stochastic Difference Volterra Equations and Some Auxiliary Statements

Here, an origin of stochastic difference Volterra equations of neutral type is described and some necessary definitions and auxiliary assertions are given.

1.1 Origin of Stochastic Difference Volterra Equations of Neutral Type

The ordinary differential equation is considered usually in the form

$$\dot{x}(t) = f(t, x(t)), \tag{1.1}$$

where $x(t)$ is a value of the solution of equation (1.1) in the moment of time t. If the right-hand side of the differential equation (1.1) depends on the whole trajectory of $x(s)$, $s \leq t$, we obtain a differential equation with delay [153, 154]

$$\dot{x}(t) = f(t, x_t), \qquad x_t = x(t+s), \quad s \leq 0. \tag{1.2}$$

Adding to the right-hand side of Eq. (1.2) stochastic perturbations of the white noise type $\dot{w}(t)$, where $w(t)$ is the standard Wiener process, we obtain Ito's stochastic differential equation [67–72]

$$\dot{x}(t) = a(t, x_t) + b(t, x_t)\dot{w}(t). \tag{1.3}$$

Equation (1.3) is usually considered in the differential form

$$\mathrm{d}x(t) = a(t, x_t)\mathrm{d}t + b(t, x_t)\mathrm{d}w(t) \tag{1.4}$$

L. Shaikhet, *Optimal Control of Stochastic Difference Volterra Equations*,
Studies in Systems, Decision and Control 17, DOI 10.1007/978-3-319-13239-6_1

or in the integral form

$$x(t) = x(0) + \int_0^t a(s, x_s)\mathrm{d}s + \int_0^t b(s, x_s)\mathrm{d}w(s). \tag{1.5}$$

The following generalization of Eq. (1.4) gives the differential equation of neutral type [105, 106, 181]

$$\mathrm{d}(x(t) - \Phi(t, x_t)) = a(t, x_t)\mathrm{d}t + b(t, x_t)\mathrm{d}w(t),$$

or similarly to (1.5) the same in the integral form

$$x(t) = \eta(0) + \Phi(t, x_t) + \int_0^t a(s, x_s)\mathrm{d}s + \int_0^t b(s, x_s)\mathrm{d}w(s),$$
$$\eta(0) = x(0) - \Phi(0, x_0). \tag{1.6}$$

Adding a dependence on t on the right-hand side of Eq. (1.6), we obtain the stochastic integral Volterra equation of neutral type [9, 109]

$$x(t) = \eta(t) + \Phi(t, x_t) + \int_0^t a(t, s, x_s)\mathrm{d}s + \int_0^t b(t, s, x_s)\mathrm{d}w(s). \tag{1.7}$$

Numerical simulation of the solution of equation (1.7) brings us a necessity to consider the difference analogue of this equation

$$x(i+1) = \eta(i+1) + \Phi(i+1, x_{i+1}) + \sum_{j=0}^{i} a(i, j, x_j) + \sum_{j=0}^{i} b(i, j, x_j)\xi(j+1). \tag{1.8}$$

The problems of optimal control for the integral equation of the type of (1.7) are considered in [9, 109]. The aim of this book is to extend the mathematical theory of optimal control on the class of stochastic difference Volterra equations of the type of (1.8).

1.2 A Necessary Condition for Optimality of Control: Statement of the Problem

Definition 1.1 ([110]) Let X and Y be two normalized spaces and F be a map from X to Y. The limit

$$F_0(x, h) = \lim_{\varepsilon \to 0} \frac{1}{\varepsilon}[F(x + \varepsilon h) - F(x)]$$

is called the Gateaux differential of the map F in the point x, where the convergence is understanding in the norm of the space Y. If the Gateaux differential $F_0(x, h)$ is linear with respect to h, i.e.,

$$F_0(x, h) = F_1(x)h,$$

where $F_1(x)$ is a linear operator, then $F_1(x)$ is called the Gateaux derivative of the functional $F(x)$ in the point x.

Let us consider the problem of optimal control $\{x_u, J(u), \mathbf{U}\}$ with a controlled process x_u, a performance functional $J(u)$ and a set of admissible controls $\mathbf{U}$. Let u_0 be the optimal control of this problem, that is, $J(u_0) = \inf_{u \in \mathbf{U}} J(u)$, and let u_ε, $\varepsilon \geq 0$, be a set of admissible controls for which there exists the finite limit

$$J_0(u_0) = \lim_{\varepsilon \to 0} \frac{1}{\varepsilon}[J(u_\varepsilon) - J(u_0)]. \tag{1.9}$$

It is clear that the inequality $J(u_0) \geq 0$ provides a necessary condition for the optimality of the control u_0. It is clear also that the existence of the limit (1.9) and the form of the optimality condition depend on the method of the controls u_ε construction. Put, for instance,

$$u_\varepsilon = u_0 + \varepsilon v, \qquad u_0, v \in \mathbf{U}. \tag{1.10}$$

Then (Definition 1.1) $J_0(u_0)$ is the Gateaux differential of the functional $J(u)$ at the point $u = u_0$, that is, $J_0(u_0) = J_0(u_0, v)$. If the Gateaux differential is linear with respect to v, then the representation

$$J_0(u_0, v) = \langle J_1(u_0), v \rangle$$

holds, where $J_1(u_0)$ is the Gateaux derivative of the functional $J(u)$ at the point $u = u_0$.

If, in additional, $\mathbf{U}$ is a convex set and u_0 is an interior point of $\mathbf{U}$, then the condition $\langle J_1(u_0), v \rangle \geq 0$ for arbitrary $v \in U$ is equivalent [194] to the condition $J_1(u_0) = 0$. If the functional $J(u)$ is a convex one on $\mathbf{U}$, then the condition $J_1(u_0) = 0$ is not a necessary condition only, but also a sufficient one for the optimality of the control u_0 [194]. The last assertion is true, for instance, for quadratic performance functionals $J(u)$.

Below, this idea is applied to the optimal control problem with a controlled process x_u given by the stochastic difference Volterra equation of neutral type.

1.3 Auxiliary Definitions and Assertions

Consider the optimal control problem $\{x_u, J(u), \mathbf{U}\}$ with a controlled process x_u, a performance criterion $J(u)$ and a set of admissible controls $\mathbf{U}$. Introduce also the auxiliary optimal control problem $\{y_u, I(u), \mathbf{U}\}$ with a controlled process y_u, a performance criterion $I(u)$ and the same set of admissible controls $\mathbf{U}$.

Put $V = \inf_{u\in\mathbf{U}} J(u)$, $W = \inf_{u\in\mathbf{U}} I(u)$ and denote by w the optimal control of the control problem $\{y_u, I(u), \mathbf{U}\}$, i.e., $W = I(w)$. Define also the closeness of two optimal control problems $\{x_u, J(u), \mathbf{U}\}$ and $\{y_u, I(u), \mathbf{U}\}$ by the function $\rho(J, I) = \sup_{u\in\mathbf{U}} |J(u) - I(u)|$.

Lemma 1.1 *Under the above conditions the following inequality is true:*

$$0 \le J(w) - V \le 2\rho(J, I). \tag{1.11}$$

Proof From $|J(u) - I(u)| \le \rho(J, I)$ it follows that $|V - W| \le \rho(J, I)$. Indeed, since

$$-\rho(J, I) \le J(u) - I(u) \le \rho(J, I),$$

we have

$$J(u) \le I(u) + \rho(J, I) \quad \text{and} \quad I(u) \le J(u) + \rho(J, I).$$

Consequently,

$$V \le W + \rho(J, I) \quad \text{and} \quad W \le V + \rho(J, I),$$

i.e., $|W - V| \le \rho(J, I)$. Thus, we obtain

$$0 \le J(w) - V \le |J(w) - I(w)| + |W - V| \le 2\rho(J, I).$$

The lemma is proved.

Lemma 1.2 *Let $z(i)$, $i = 0, 1, \dots, N - 1$, be a nonnegative nondecreasing sequence, i.e.,*

$$0 \le z(0) \le z(1) \le \cdots \le z(N - 1), \tag{1.12}$$

and $y(i)$ be a nonnegative sequence that satisfies the condition

$$y(i + 1) \le C\left[z(i) + \sum_{j=1}^{i} y(j)\right],$$
$$0 \le i \le N - 1, \qquad C > 0. \tag{1.13}$$

Then

$$y(i+1) \leq C_1 z(i), \qquad i = 0, 1, \ldots, N-1,$$
$$C_1 = C(1+C)^{N-1}. \tag{1.14}$$

Proof Note that for $i = 0$ from (1.13) we have $y(1) \leq Cz(0)$. Assume that

$$y(j) \leq C(1+C)^{j-1} z(j-1), \qquad j = 1, \ldots, i, \tag{1.15}$$

and prove that (1.15) holds for $j = i+1$. Really, via (1.12) $z(j-1) \leq z(i)$ for $j \leq i$. From this and (1.13), (1.15) it follows that for $i \leq N-1$

$$\begin{aligned} y(i+1) &\leq C\left[z(i) + \sum_{j=1}^{i} C(1+C)^{j-1} z(j-1)\right] \\ &\leq C\left[1 + C\frac{(1+C)^i - 1}{C}\right] z(i) \\ &= C(1+C)^i z(i) \\ &\leq C_1 z(i). \end{aligned}$$

The proof is completed.

Definition 1.2 The matrix $R(i,j)$ is called the resolvent of a kernel $a(i,j)$ if the solution of the equation

$$x(i+1) = \eta(i+1) + \sum_{j=m}^{i} a(i,j)x(j),$$
$$m \leq i \leq l, \qquad x(m) = \eta(m), \tag{1.16}$$

is represented in the form

$$x(i+1) = \eta(i+1) + \sum_{j=m}^{i} R(i,j)\eta(j),$$
$$m \leq i \leq l, \qquad x(m) = \eta(m). \tag{1.17}$$

Lemma 1.3 *Let the solution of equation* (1.16) *be represented in the form* (1.17). *Then the resolvent* $R(i,j)$ *is defined by the recurrent formulas*

$$R(i,j) = a(i,j) + \sum_{l=j+1}^{i} a(i,l)R(l-1,j), \qquad 0 \leq j \leq i, \tag{1.18}$$

$$R(i,j)=a(i,j)+\sum_{l=j+1}^{i}R(i,l)a(l-1,j),\qquad 0\le j\le i. \tag{1.19}$$

Proof Substituting (1.17) into (1.16), we obtain

$$\eta(i+1)+\sum_{j=m}^{i}R(i,j)\eta(j)=\eta(i+1)+\sum_{j=m}^{i}a(i,j)\left[\eta(j)+\sum_{l=m}^{j-1}R(j-1,l)\eta(l)\right]$$

or

$$\sum_{j=m}^{i}R(i,j)\eta(j)=\sum_{j=m}^{i}a(i,j)\eta(j)+\sum_{j=m}^{i}\left[\sum_{l=j+1}^{i}a(i,l)R(l-1,j)\right]\eta(j)$$

that is equivalent to (1.18). Substituting $\eta(i)$ from (1.16) into (1.17), we obtain

$$x(i+1)=x(i+1)-\sum_{j=m}^{i}a(i,j)x(j)+\sum_{j=m}^{i}R(i,j)\left[x(j)-\sum_{l=m}^{j-1}R(j-1,l)x(l)\right]$$

or

$$\sum_{j=m}^{i}R(i,j)x(j)=\sum_{j=m}^{i}a(i,j)x(j)+\sum_{j=m}^{i}\left[\sum_{l=j+1}^{i}R(i,l)a(l-1,j)\right]x(j)$$

that is equivalent to (1.19). The proof is completed.

Corollary 1.1 *Suppose the kernel $a(i,j)$ satisfies the condition $a(i,j)=a(i-j)$. Then the resolvent $R(i,j)$ satisfies the similar condition $R(i,j)=R(i-j)$ and the recurrent formulas* (1.18) *and* (1.19)*, respectively, take the forms*

$$R(i)=a(i)+\sum_{j=0}^{i-1}a(i-1-j)R(j) \tag{1.20}$$

and

$$R(i)=a(i)+\sum_{j=0}^{i-1}R(i-1-j)a(j). \tag{1.21}$$

Definition 1.3 The matrix $R(i,j)$ is called the Fredholm resolvent of a kernel $a(i,j)$ if the solution of the equation

$$x(i) = \eta(i) + \sum_{j=m}^{l} a(i, j)x(j),$$
$$m \leq i, j \leq l \leq \infty, \tag{1.22}$$

is represented in the form

$$x(i) = \eta(i) + \sum_{j=m}^{l} R(i, j)\eta(j),$$
$$m \leq i, j \leq l \leq \infty. \tag{1.23}$$

Lemma 1.4 *Let the solution of equation* (1.22) *be represented in the form* (1.23). *Then the Fredholm resolvent* $R(i, j)$ *is defined by the recurrent formulas*

$$R(i, j) = a(i, j) + \sum_{k=m}^{l} a(i, k)R(k, j). \tag{1.24}$$

$$R(i, j) = a(i, j) + \sum_{k=m}^{l} R(i, k)a(k, j), \tag{1.25}$$

Proof Substituting (1.23) into (1.22), we obtain

$$\eta(i) + \sum_{j=m}^{l} R(i, j)\eta(j) = \eta(i) + \sum_{k=m}^{l} a(i, k)\left[\eta(k) + \sum_{j=m}^{l} R(k, j)\eta(j)\right]$$

or

$$\sum_{j=m}^{l} R(i, j)\eta(j) = \sum_{j=m}^{l} a(i, j)\eta(j) + \sum_{j=m}^{l}\sum_{k=m}^{l} a(i, k)R(k, j)\eta(j)$$

that is equivalent to (1.24). Substituting $\eta(i)$ from (1.22) into (1.23), we obtain

$$x(i) = x(i) - \sum_{j=m}^{l} a(i, j)x(j) + \sum_{j=m}^{l} R(i, j)\left[x(j) - \sum_{k=m}^{l} a(j, k)x(k)\right]$$

or

$$\sum_{j=m}^{l} R(i, j)x(j) = \sum_{j=m}^{l} a(i, j)x(j) + \sum_{j=m}^{l}\sum_{k=m}^{l} R(i, k)a(k, j)x(j)$$

that is equivalent to (1.25). The proof is completed.

Lemma 1.5 *Let $z(i)$ be a solution of the equation*

$$z(i+1) = \eta(i+1) + \sum_{j=0}^{i} P(i,j)z(j), \qquad z(0) = \eta(0), \tag{1.26}$$

where $\eta(i) \geq 0$, $P(i,j) \geq 0$, $0 \leq j \leq i$, and $y(i)$ be a nonnegative sequence that satisfies the condition

$$y(i+1) \leq \eta(i+1) + \sum_{j=0}^{i} P(i,j)y(j), \qquad y(0) = \eta(0). \tag{1.27}$$

Then the condition $y(i) \leq z(i)$, $i = 0, 1, \ldots$, holds.

Proof Let us construct the sequence $y_k(i)$, $k = 0, 1, \ldots$, in the following way:

$$\begin{aligned} y_{k+1}(i+1) &= \eta(i+1) + \sum_{j=0}^{i} P(i,j)y_k(j), \\ y_k(0) &= \eta(0), \quad k = 0, 1, \ldots, \\ y_0(i) &= y(i), \quad i = 0, 1, \ldots. \end{aligned} \tag{1.28}$$

For $k = 0$ from (1.28), (1.27) it follows that

$$\begin{aligned} y_1(i+1) &= \eta(i+1) + \sum_{j=0}^{i} P(i,j)y_0(j) \\ &= \eta(i+1) + \sum_{j=0}^{i} P(i,j)y(j) \\ &\geq y(i+1) = y_0(i+1). \end{aligned}$$

Assume that $y_k(j) \geq y_{k-1}(j)$ for $k \geq 1$ and prove that this condition holds for $k+1$. Really, via (1.28) we have

$$\begin{aligned} y_{k+1}(i+1) &= \eta(i+1) + \sum_{j=0}^{i} P(i,j)y_k(j) \\ &\geq \eta(i+1) + \sum_{j=0}^{i} P(i,j)y_{k-1}(j) \\ &= y_k(i+1). \end{aligned}$$

Thus, from this it follows that for each $i = 0, 1, \ldots$

$$y(i) = y_0(i) \le y_k(i) \le y_{k+1}(i),$$

i.e., the sequence $y_k(i)$, $k = 0, 1, \ldots$, do not decrease and therefore has a limit

$$\lim_{k\to\infty} y_k(i) = z(i)$$

that via (1.28), (1.26) is a solution of equation (1.26). The proof is completed.

Definition 1.4 The matrix $A^{\oplus}$ of dimension $n \times m$ is called the pseudoinverse matrix to the given $m \times n$-matrix A, if the following conditions hold:

$$AA^{\oplus}A = A, \qquad A^{\oplus} = PA' = A'Q, \tag{1.29}$$

where A' is the transposed matrix, P and Q are some matrices.

Remark 1.1 The second condition (1.29) means that rows and columns of the matrix $A^{\oplus}$ are linear combinations of rows and columns, respectively, of the matrix A'.

Lemma 1.6 ([130]) *The matrix $A^{\oplus}$ that satisfies the conditions* (1.29) *there exists a unique one.*

Definition 1.5 The random vector $\xi = (\xi_1, \ldots, \xi_n)$ ia called Gaussian (or normal) vector if his characteristic function

$$\varphi_\xi(z) = \mathbf{E}e^{iz'\xi}, \qquad z = (z_1, \ldots, z_n),$$

is defined by the formula

$$\varphi_\xi(z) = e^{iz'm - \frac{1}{2}z'Rz},$$

where

$$m = \mathbf{E}\xi, \qquad R = \mathrm{cov}(\xi, \xi) = \mathbf{E}(\xi - m)(\xi - m)'.$$

Theorem 1.1 (about normal correlation) ([130]) *Let* $(\theta, \xi) = ([\theta_1, \ldots, \theta_k], [\xi_1, \ldots, \xi_l])$ *be the Gaussian vector with*

$$m_\theta = \mathbf{E}\theta, \qquad m_\xi = \mathbf{E}\xi,$$
$$\mathbf{D}_{\theta\theta} = \mathrm{cov}(\theta, \theta), \quad \mathbf{D}_{\theta\xi} = \mathrm{cov}(\theta, \xi), \quad \mathbf{D}_{\xi\xi} = \mathrm{cov}(\xi, \xi).$$

Then the conditional mathematical expectation $\mathbf{E}(\theta/\xi)$ *and the conditional covariation*

$$\mathrm{cov}(\theta, \theta/\xi) = \mathbf{E}([\theta - \mathbf{E}(\theta/\xi)][\theta - \mathbf{E}(\theta/\xi)]'/\xi)$$

are defined by the formulas

$$\mathbf{E}(\theta/\xi) = m_\theta + \mathbf{D}_{\theta\xi}\mathbf{D}_{\xi\xi}^{\oplus}(\xi - m_\xi),$$
$$\operatorname{cov}(\theta, \theta/\xi) = \mathbf{D}_{\theta\theta} - \mathbf{D}_{\theta\xi}\mathbf{D}_{\xi\xi}^{\oplus}(\mathbf{D}_{\theta\xi})'.$$

Corollary 1.2 *Let the random variables* $(\theta, \xi_1, \ldots, \xi_l)$ *be the Gaussian vector,* $\xi_1, \ldots, \xi_l$ *be mutually independent and* $\mathbf{D}\xi_i > 0$, $i = 1, \ldots, l$. *Then*

$$\mathbf{E}(\theta/\xi_1, \ldots, \xi_l) = \mathbf{E}\theta + \sum_{i=1}^{l} \frac{\operatorname{cov}(\theta, \xi_i)}{\mathbf{D}\xi_i}(\xi_i - \mathbf{E}\xi_i).$$

In particular, if $\mathbf{E}\theta = \mathbf{E}\xi_i = 0$ then

$$\mathbf{E}(\theta/\xi_1, \ldots, \xi_l) = \sum_{i=1}^{l} \frac{\operatorname{cov}(\theta, \xi_i)}{\mathbf{D}\xi_i}\xi_i.$$

1.4 Some Useful Lemmas

In this section some simple statements are included that will be used below.

Lemma 1.7 *If* $a_i \geq 0$, $p \geq 1$ *then*

$$\left(\sum_{i=1}^{N} a_i\right)^p \leq N^{p-1}\sum_{i=1}^{N} a_i^p. \tag{1.30}$$

Proof simply follows from the Cauchy-Bunyakovskii inequality.

Lemma 1.8 *For arbitrary* $a \geq 0$, $b \geq 0$, $\alpha > 0$

$$(a+b)^2 \leq (1+\alpha)a^2 + \left(1 + \frac{1}{\alpha}\right)b^2. \tag{1.31}$$

For arbitrary $a \geq 0$, $b \geq 0$, $c \geq 0$, $\alpha_1 > 0$, $\alpha_2 > 0$, $\alpha_3 > 0$

$$\begin{aligned}(a+b+c)^2 &\leq (1+\alpha_1+\alpha_2)\,a^2 \\ &\quad + \left(1+\alpha_3+\frac{1}{\alpha_1}\right)b^2 + \left(1+\frac{1}{\alpha_2}+\frac{1}{\alpha_3}\right)c^2.\end{aligned} \tag{1.32}$$

Proof It is easy to check that the minimum of the right-hand side of the inequality (1.32) with respect to positive $\alpha_1, \alpha_2, \alpha_3$ coincides with the left-hand side of (1.32).

Putting in (1.32) $c = 0$ and then $\alpha_1 = \alpha$, $\alpha_2 = \alpha_3 = 0$, one can obtain (1.31). The lemma is proven.

Lemma 1.9 *A number of combinations from a given set of n elements*

$$\binom{n}{m} = \frac{n!}{m!(n-m)!}, \qquad m = 0, 1, \dots, n, \tag{1.33}$$

satisfies the following equality

$$\binom{n}{m} = \sum_{j=m-1}^{n-1} \binom{j}{m-1}, \qquad m = 1, \dots, n. \tag{1.34}$$

Proof Using the mathematical induction method, note that for $n = 1$ the equality (1.34) holds trivial. Suppose it holds for some $n \geq 1$ and prove that it holds for $n+1$. Really, via (1.34), (1.33) we obtain

$$\begin{aligned}
\binom{n+1}{m} &= \sum_{j=m-1}^{n} \binom{j}{m-1} \\
&= \sum_{j=m-1}^{n-1} \binom{j}{m-1} + \binom{n}{m-1} \\
&= \binom{n}{m} + \binom{n}{m-1} \\
&= \frac{n!}{m!(n-m)!} + \frac{n!}{(m-1)!(n-m+1)!} \\
&= \frac{n!}{m!(n-m+1)!}(n-m+1+m) \\
&= \frac{(n+1)!}{m!(n+1-m)!}.
\end{aligned}$$

The proof is completed.

Chapter 2
Optimal Control

In this chapter, optimal control problems for nonlinear and linear equations are considered. Necessary condition for optimality of control for nonlinear control problem is obtained and it is used for construction of the synthesis of optimal control for a linear-quadratic problem.

2.1 A Necessary Condition for Optimality of Control for Nonlinear System

2.1.1 Statement of the Problem

Let $\{\Omega, \mathfrak{F}, \mathbf{P}\}$ be a basic probability space, $\mathbf{E}$ be an expectation, $Z_0 = \{-h, -h+1, \ldots, -1, 0\}$, $h \in [0, \infty]$, and $Z = \{0, 1, \ldots, N\}$ be a discrete time, $\mathfrak{F}_i \subset \mathfrak{F}$, $i \in Z$, be a family of σ-algebras, $\mathbf{E}_i = \mathbf{E}\{\cdot/\mathfrak{F}_i\}$ be a conditional expectation, H_0 be a space of $\mathfrak{F}_0$-adapted random functions $\varphi(j) \in \mathbf{R}^n$, $j \in Z_0$, such that

$$\|\varphi\|^2 = \max_{j \in Z_0} \mathbf{E}|\varphi(j)|^2 < \infty, \tag{2.1}$$

H be a space of $\mathfrak{F}_i$-adapted random functions $x(i) \in \mathbf{R}^n$, $i \in Z$, such that

$$\|x\|_i^2 = \max_{0 \le j \le i} \mathbf{E}|x(j)|^2 < \infty. \tag{2.2}$$

Consider the nonlinear controlled system

$$\begin{aligned} x(i+1) &= \eta(i+1) + \Phi(i+1, x_{i+1}) \\ &\quad + \sum_{j=0}^{i} a(i, j, x_j, u(j)) + \sum_{j=0}^{i} b(i, j, x_j, u(j))\xi(j+1), \\ i &= 0, 1, \ldots, N-1, \quad x(j) = \varphi_0(j), \quad j \in Z_0, \quad \varphi_0 \in H_0, \end{aligned} \tag{2.3}$$

L. Shaikhet, *Optimal Control of Stochastic Difference Volterra Equations*,
Studies in Systems, Decision and Control 17, DOI 10.1007/978-3-319-13239-6_2

with the performance functional

$$J(u) = \mathbf{E}\left[F(x_N) + \sum_{j=0}^{N-1} G(j, x_j, u(j))\right]. \tag{2.4}$$

Here, $x(i)$ is a value of the process x in the time moment i, x_i is a trajectory of $x(j)$ from the moment of time $j = -h$ until the moment of time $i \in Z$, $\eta \in H$, $\xi(i) \in \mathbf{R}^l$ is a sequence of $\mathfrak{F}_i$-adapted Gaussian random variables, that are mutually independent and do not depend on $\eta(i)$ such that $\mathbf{E}\xi(i) = 0, \mathbf{E}\xi(i)\xi'(i) = I, i \in Z, I$ is the identical matrix.

Definition 2.1 Arbitrary $\mathfrak{F}_i$-adapted function $u(i) \in \mathbf{R}^m$, $i \in Z$, for which there exists the solution of the equation (2.3) and the performance functional (2.4) is bounded, is called an admissible control, $\mathbf{U}$ is the set of admissible controls.

Our nearest goal is to calculate the limit (1.9), (1.10) for the optimal control problem (2.3), (2.4). For this goal, we will assume that the functionals $\Phi(i, \varphi) \in \mathbf{R}^n$, $a(i, j, \varphi, u) \in \mathbf{R}^n$, $F(\varphi) \in \mathbf{R}^1$, $G(j, \varphi, u) \in \mathbf{R}^1$, and $n \times l$-matrix $b(i, j, \varphi, u)$ have Gateaux derivatives with respect to φ denoted hereafter by ∇ and the functionals $a(i, j, \varphi, u)$, $b(i, j, \varphi, u)$, and $G(j, \varphi, u)$, $0 \le j \le i \le N$ have derivatives with respect to u denoted hereafter by ∇_u, $u \in \mathbf{R}^m$, $\varphi \in \tilde{H}$, where $\tilde{H}$ is a space of the functions $\varphi(i) \in \mathbf{R}^n$, $i \in Z_0 \cup Z$.

Let X and Y be two normed spaces and $f(x)$ be a map of X to Y. Then the Gateaux derivative $\nabla f(\varphi_0)$ of this map for a fixed $\varphi_0 \in \tilde{H}$ is [110] a linear operator that maps X to Y. If $Y = \mathbf{R}^1$ then $\langle \nabla f(\varphi_0), \varphi \rangle$ is the value of the linear functional $\nabla f(\varphi_0)$ on the element $\varphi \in \tilde{H}$.

Also we assume that the functional $\Phi(i, \varphi)$ depends on the values of the function $\varphi(j)$ for $j = -h, -h+1, \dots, i$, the functionals $a(i, j, \varphi, u)$, $b(i, j, \varphi, u)$ depend on the values of the function $\varphi(l)$ for $l = -h, -h+1, \dots, j$, $\varphi \in \tilde{H}$ and satisfy the conditions

$$|\Phi(i, \varphi)| \le \sum_{j=-h}^{i} (1 + |\varphi(j)|) K_0(j), \tag{2.5}$$

$$|a(i, j, \varphi, u)|^2 + |b(i, j, \varphi, u)|^2 \le \sum_{l=-h}^{j} (1 + |u|^2 + |\varphi(l)|^2) K_1(l), \tag{2.6}$$

for arbitrary functions $\varphi_1, \varphi_2 \in \tilde{H}$

$$|\Phi(i, \varphi_1) - \Phi(i, \varphi_2)| \le \sum_{j=-h}^{i} |\varphi_1(j) - \varphi_2(j)| K_0(j), \tag{2.7}$$

$$|a(i,j,\varphi_1,u)-a(i,j,\varphi_2,u)|^2+|b(i,j,\varphi_1,u)-b(i,j,\varphi_2,u)|^2 \leq \sum_{l=-h}^{j}|\varphi_1(l)-\varphi_2(l)|^2K_1(l), \tag{2.8}$$

$$|\nabla\Phi(i,\varphi_1)\varphi| \leq \sum_{j=0}^{i}|\varphi(j)|K_0(j), \tag{2.9}$$

$$|\nabla a(i,j,\varphi_1,u)\varphi|^2+|\nabla b(i,j,\varphi_1,u)\varphi|^2 \leq \sum_{l=0}^{j}|\varphi(l)|^2K_1(l), \tag{2.10}$$

$$|\nabla_u a(i,j,\varphi,u)|+|\nabla_u b(i,j,\varphi,u)| \leq C, \tag{2.11}$$

$$|(\nabla\Phi(i,\varphi_1)-\nabla\Phi(i,\varphi_2))\varphi|^2 \leq \sum_{j=-h}^{i}|\varphi_1(j)-\varphi_2(j)|^2|\varphi(j)|^2K_0(j). \tag{2.12}$$

Here and everywhere below, it is supposed that

$$K_0=\max_{-h\leq i\leq N}K_0(i)<1, \qquad K_1=\max_{-h\leq i\leq N}K_1(i)<\infty,$$
$$\text{if } h=\infty \text{ then } \sum_{j=-h}^{N}K_l(j)<\infty, \quad l=0,1. \tag{2.13}$$

For arbitrary $u_1,u_2\in\mathbf{U}$

$$\begin{aligned}&(\nabla a(i,j,\varphi_1,u_1)-\nabla a(i,j,\varphi_2,u_2))\varphi|^2\\&\quad+|(\nabla b(i,j,\varphi_1,u_1)-\nabla b(i,j,\varphi_2,u_2))\varphi|^2\\&\leq\sum_{l=0}^{j}\left(|\varphi_1(l)-\varphi_2(l)|^2+|u_1-u_2|^2\right)|\varphi(l)|^2K_1(l),\end{aligned} \tag{2.14}$$

$$\begin{aligned}&|\nabla_u a(i,j,\varphi,u_1)-\nabla_u a(i,j,\varphi,u_2)|\\&\quad+|\nabla_u b(i,j,\varphi,u_1)-\nabla_u b(i,j,\varphi,u_2)|\\&\leq C|u_1-u_2|.\end{aligned} \tag{2.15}$$

Regarding the performance functional we assume that $F(\varphi)$ depends on the values of the function $\varphi(j)$ for $j=-h,-h+1,\ldots,N$, $G(i,\varphi,u)$ depends on the values of the function $\varphi(j)$ for $j = -h, -h+1, \dots, i$, $\varphi \in \tilde{H}$, and both these functionals satisfy the conditions

$$|F(\varphi)| \leq \sum_{j=-h}^{N} \left(1 + |\varphi(j)|^2\right) K_1(j), \tag{2.16}$$

$$|G(i, \varphi, u)| \leq \sum_{j=-h}^{i} \left(1 + |u|^2 + |\varphi(j)|^2\right) K_1(j), \tag{2.17}$$

for arbitrary functions $\varphi_1, \varphi_2 \in \tilde{H}$

$$|\langle \nabla F(\varphi_1), \varphi \rangle| \leq \sum_{j=0}^{N} (1 + |\varphi_1(j)|)\, |\varphi(j)| K_1(j), \tag{2.18}$$

$$|\langle \nabla G(i, \varphi_1, u), \varphi \rangle| \leq \sum_{j=0}^{i} (1 + |u| + |\varphi_1(j)|)\, |\varphi(j)| K_1(j), \tag{2.19}$$

$$|\langle \nabla F(\varphi_1) - \nabla F(\varphi_2), \varphi \rangle| \leq \sum_{j=0}^{N} |\varphi_1(j) - \varphi_2(j)| |\varphi(j)| K_1(j), \tag{2.20}$$

for arbitrary $\varphi_1, \varphi_2 \in \tilde{H}$, and $u_1, u_2 \in \mathbf{U}$

$$|\langle \nabla G(i, \varphi_1, u_1) - \nabla G(i, \varphi_2, u_2), \varphi \rangle|$$
$$\leq \sum_{j=0}^{i} (|\varphi_1(j) - \varphi_2(j)| + |u_1 - u_2|)\, |\varphi(j)| K_1(j), \tag{2.21}$$

$$|\nabla_u G(i, \varphi, u)| \leq \sum_{j=0}^{i} (1 + |u| + |\varphi(j)|)\, K_1(j), \tag{2.22}$$

$$|\nabla_u G(i, \varphi, u_1) - \nabla_u G(i, \varphi, u_2)| \leq C |u_1 - u_2|. \tag{2.23}$$

2.1.2 Auxiliary Assertions

In order to calculate the limit (1.9), (1.10) for the optimal control problem (2.3), (2.4) we need the following auxiliary assertions.

Lemma 2.1 *Let the conditions* (2.5), (2.6), (2.13), (2.16), (2.17) *hold and* $u \in H$, *i.e., the control* $u(i)$ *satisfies the condition*

$$\|u\|_i^2 = \max_{0 \leq j \leq i} \mathbf{E}|u(j)|^2 < \infty, \quad i \in Z. \tag{2.24}$$

Then the solution $x(i)$ *of the Eq.* (2.3) *satisfies the condition* (2.2) *and the performance functional* (2.4) *is bounded.*

Proof From the Eq. (2.3), we have

$$|x(i+1)| \leq |\eta(i+1)| + |\Phi(i+1, x_{i+1})| + \sum_{j=0}^{i} |a(i,j,x_j,u(j))| + \left|\sum_{j=0}^{i} b(i,j,x_j,u(j))\xi(j+1)\right|, \quad i = 0, 1, \ldots, N-1. \tag{2.25}$$

Via (2.5), (2.13) we obtain

$$\begin{aligned} |\Phi(i+1, x_{i+1})| &\leq \sum_{j=-h}^{i+1} (1 + |x(j)|) K_0(j) \\ &\leq \sum_{j=-h}^{i+1} K_0(j) + \sum_{j=-h}^{0} |\varphi_0(j)| K_0(j) \\ &\quad + \sum_{j=1}^{i} |x(j)| K_0(j) + |x(i+1)| K_0. \end{aligned} \tag{2.26}$$

From (2.25), (2.26) via (2.13), it follows that

$$\begin{aligned} 0 &\leq (1 - K_0)|x(i+1)| \\ &\leq |\eta(i+1)| + \sum_{j=-h}^{i+1} K_0(j) + \sum_{j=-h}^{0} |\varphi_0(j)| K_0(j) + \sum_{j=1}^{i} |x(j)| K_0(j) \\ &\quad + \sum_{j=0}^{i} |a(i,j,x_j,u(j))| + \left|\sum_{j=0}^{i} b(i,j,x_j,u(j))\xi(j+1)\right|. \end{aligned}$$

Squaring the obtained inequality, calculating the mathematical expectation, and using the properties of the process $\xi(j)$, via (1.30), we obtain

$$\begin{aligned} (1-K_0)^2 \mathbf{E}|x(i+1)|^2 \leq 6 \Bigg[& \mathbf{E}|\eta(i+1)|^2 + \left(\sum_{j=-h}^{i+1} K_0(j) \right)^2 \\ & + \sum_{l=-h}^{0} K_0(l) \sum_{j=-h}^{0} \mathbf{E}|\varphi_0(j)|^2 K_0(j) \end{aligned}$$

$$+\sum_{l=1}^{i} K_0(l) \sum_{j=1}^{i} \mathbf{E}|x(j)|^2 K_0(j)$$

$$\left. + N \sum_{j=0}^{i} \left(\mathbf{E}|a(i, j, x_j, u(j))|^2 + \mathbf{E}|b(i, j, x_j, u(j))|^2 \right) \right] .$$

Note that $\varphi_0 \in H_0$, i.e., φ_0 satisfies the condition (2.1). So, via (2.6), (2.13) we have

$$\begin{aligned}
&\sum_{j=0}^{i} \left(\mathbf{E}|a(i, j, x_j, u(j))|^2 + \mathbf{E}|b(i, j, x_j, u(j))|^2 \right) \\
&\quad \le \sum_{j=0}^{i} \sum_{l=-h}^{j} (1 + \mathbf{E}|u(j)|^2 + \mathbf{E}|x(l)|^2) K_1(l) \\
&\quad = \sum_{j=0}^{i} (1 + \mathbf{E}|u(j)|^2) \sum_{l=-h}^{j} K_1(l) \\
&\qquad + \sum_{j=0}^{i} \left(\sum_{l=-h}^{0} \mathbf{E}|\varphi_0(l)|^2 K_1(l) + \sum_{l=1}^{j} \mathbf{E}|x(l)|^2 K_1(l) \right) \\
&\quad \le \sum_{j=0}^{i} \left((1 + \mathbf{E}|u(j)|^2) \sum_{l=-h}^{j} K_1(l) + \|\varphi_0\|^2 \sum_{l=-h}^{0} K_1(l) \right) \\
&\qquad + N K_1 \sum_{l=1}^{i} \mathbf{E}|x(l)|^2.
\end{aligned}$$

As a result, using (2.13), we obtain

$$\begin{aligned}
&\mathbf{E}|x(i+1)|^2 \\
&\quad \le \frac{6}{(1-K_0)^2} \left[\mathbf{E}|\eta(i+1)|^2 + \left(\sum_{j=-h}^{i+1} K_0(j) \right)^2 + N^2 K_1 \sum_{l=1}^{i} \mathbf{E}|x(l)|^2 \right. \\
&\qquad + \|\varphi_0\|^2 \left(\sum_{l=-h}^{0} K_0(l) \right)^2 + K_0 \sum_{l=1}^{N} K_0(l) \sum_{j=1}^{i} \mathbf{E}|x(j)|^2 \\
&\qquad \left. + N \sum_{j=0}^{i} \left((1 + \mathbf{E}|u(j)|^2) \sum_{l=-h}^{j} K_1(l) + \|\varphi_0\|^2 \sum_{l=-h}^{0} K_1(l) \right) \right] .
\end{aligned}$$

From this, it follows that

$$\mathbf{E}|x(i+1)|^2 \leq C\left(z(i) + \sum_{j=1}^{i} \mathbf{E}|x(j)|^2\right),$$

where

$$C = \frac{6}{(1-K_0)^2}\left(K_0 \sum_{l=1}^{N} K_0(l) + K_1 N^2\right),$$

$$z(i) = \left(K_0 \sum_{l=1}^{N} K_0(l) + K_1 N^2\right)^{-1}$$
$$\times \left[\|\eta\|^2 + \|\varphi_0\|^2 \left(\sum_{l=-h}^{0} K_1(l)\right)^2 + \left(\sum_{l=-h}^{i+1} K_0(l)\right)^2\right.$$
$$\left. + N \sum_{j=0}^{i} \left(\|\varphi_0\|^2 \sum_{l=-h}^{0} K_1(l) + (1 + \mathbf{E}|u(j)|^2) \sum_{l=-h}^{j} K_1(l)\right)\right].$$

Via η, $u \in H$, $\varphi_0 \in H_0$, and (2.13), we have that $z(i)$ is a nonnegative nondecreasing sequence and $\max_{0 \leq i \leq N} z(i) < \infty$. So, via Lemma 1.2 and (1.14) there exists $C_1 > 0$ such that

$$\mathbf{E}|x(i+1)|^2 \leq C_1 z(i) < \infty, \qquad i = 0, 1, \ldots, N-1, \tag{2.27}$$

and $x(i)$ satisfies (2.2).

For the performance functional (2.4) via (2.16), (2.17) we have

$$J(u) \leq \sum_{j=-h}^{N} (1 + \mathbf{E}|x(j)|^2) K_1(j)$$
$$+ \sum_{j=0}^{N-1} \sum_{l=-h}^{j} (1 + \mathbf{E}|u(j)|^2 + \mathbf{E}|x(l)|^2) K_1(l).$$

From this, it follows that

$$\begin{aligned} J(u) \leq & \sum_{j=-h}^{N} K_1(j) + \sum_{j=-h}^{0} \mathbf{E}|\varphi_0(j)|^2 K_1(j) + \sum_{j=1}^{N} \mathbf{E}|x(j)|^2 K_1(j) \\ & + \sum_{j=0}^{N-1} (1 + |u(j)|^2) \sum_{l=-h}^{N-1} K_1(l) \\ & + \sum_{j=0}^{N-1} \left(\sum_{l=-h}^{0} \mathbf{E}|\varphi_0(l)|^2 K_1(l) + \sum_{l=1}^{j} \mathbf{E}|x(l)|^2 K_1(l) \right). \end{aligned}$$

As a result via (2.1), (2.2), (2.13), (2.24), (2.27) we obtain

$$\begin{aligned} J(u) & \leq (N+1)(1 + \|u\|_N^2 + \|\varphi_0\|^2 + \|x\|_N^2) \sum_{j=-h}^{N} K_1(j) \\ & < \infty. \end{aligned}$$

The proof is completed.

Let us investigate now the behavior of the process

$$q_\varepsilon(i) = \frac{1}{\varepsilon} (x_\varepsilon(i) - x_0(i)), \tag{2.28}$$

where $x_\varepsilon(i)$ is the solution of the equation (2.3) under the control

$$u_\varepsilon(i) = u_0(i) + \varepsilon v(i), \qquad \varepsilon \geq 0. \tag{2.29}$$

Definition 2.2 The process $q_\varepsilon(i)$ is called uniformly p-bounded, $p > 0$, with respect to $\varepsilon \geq 0$ if there exists $C > 0$ that does not depend on ε and such that

$$\max_{i \in Z} \mathbf{E}|q_\varepsilon(i)|^p \leq C.$$

If $p = 2$, then the process $q_\varepsilon(i)$ is called uniformly mean square bounded.

Lemma 2.2 *Let the conditions* (2.9)–(2.11), (2.13) *hold and* $\max_{i \in Z} \mathbf{E}|v(j)|^p < \infty$, $p \geq 2$. *Then q_ε is uniformly p-bounded with respect to* $\varepsilon \geq 0$.

Proof From the Eq. (2.3), it follows that

$$\begin{aligned} q_\varepsilon(i+1) &= \frac{1}{\varepsilon}\big[\Phi(i+1, x_{\varepsilon,i+1}) - \Phi(i+1, x_{0,i+1})\big] \\ &+ \frac{1}{\varepsilon}\sum_{j=0}^{i}\big[a(i,j,x_{\varepsilon j},u_\varepsilon(j)) - a(i,j,x_{0j},u_0(j))\big] \\ &+ \frac{1}{\varepsilon}\sum_{j=0}^{i}\big[b(i,j,x_{\varepsilon j},u_\varepsilon(j)) - b(i,j,x_{0j},u_0(j))\big]\xi(j+1). \end{aligned} \tag{2.30}$$

In order to transform the right-hand side of (2.30), put

$$\begin{aligned} \lambda_\varepsilon^\tau(i) &= x_0(i) + \tau\varepsilon q_\varepsilon(i), \quad u_\varepsilon^\tau(i) = u_0(i) + \tau\varepsilon v(i), \\ \Phi_\varepsilon(i) &= \int_0^1 \nabla\Phi(i, \lambda_{\varepsilon i}^\tau)\mathrm{d}\tau, \\ A_\varepsilon(i,j) &= \int_0^1 \nabla a(i,j,\lambda_{\varepsilon j}^\tau, u_\varepsilon(j))\mathrm{d}\tau, \\ a_\varepsilon(i,j) &= \int_0^1 \nabla_u a(i,j,x_{0j}, u_\varepsilon^\tau(j))\mathrm{d}\tau, \\ B_\varepsilon(i,j) &= \int_0^1 \nabla b(i,j,\lambda_{\varepsilon j}^\tau, u_\varepsilon(j))\mathrm{d}\tau, \\ b_\varepsilon(i,j) &= \int_0^1 \nabla_u b(i,j,x_{0j}, u_\varepsilon^\tau(j))\mathrm{d}\tau. \end{aligned} \tag{2.31}$$

Consider also the function

$$f(\tau) = \frac{1}{\varepsilon}\Phi(i, \lambda_{\varepsilon i}^\tau) = \frac{1}{\varepsilon}\Phi(i, x_{0i} + \tau\varepsilon q_{\varepsilon i}), \qquad \tau \in [0,1], \tag{2.32}$$

and note that via (2.32), (2.28)

$$\begin{aligned} f'(\tau) &= \nabla\Phi(i, \lambda_{\varepsilon i}^\tau)q_{\varepsilon i}, \\ \int_0^1 f'(\tau)\mathrm{d}\tau &= f(1) - f(0) = \frac{1}{\varepsilon}[\Phi(i, x_{\varepsilon i}) - \Phi(i, x_{0i})]. \end{aligned}$$

From this and (2.31), it follows that the first summand in the right-hand side of (2.30) can be represented as follows:

$$\frac{1}{\varepsilon}[\Phi(i, x_{\varepsilon i}) - \Phi(i, x_{0i})] = \int_0^1 \nabla\Phi(i, \lambda_{\varepsilon i}^{\tau})q_{\varepsilon i}\mathrm{d}\tau = \Phi_\varepsilon(i)q_{\varepsilon i}. \tag{2.33}$$

Transform now the second summand in (2.30) by the following way:

$$\begin{aligned}
&\frac{1}{\varepsilon}[a(i, j, x_{\varepsilon j}, u_\varepsilon(j)) - a(i, j, x_{0j}, u_0(j))] \\
&\quad = \frac{1}{\varepsilon}[a(i, j, x_{\varepsilon j}, u_\varepsilon(j)) - a(i, j, x_{0j}, u_\varepsilon(j))] \\
&\quad + \frac{1}{\varepsilon}[a(i, j, x_{0j}, u_\varepsilon(j)) - a(i, j, x_{0j}, u_0(j))].
\end{aligned}$$

Similarly to (2.33) via (2.31), we obtain

$$\frac{1}{\varepsilon}[a(i, j, x_{\varepsilon j}, u_\varepsilon(j)) - a(i, j, x_{0j}, u_\varepsilon(j))] = A_\varepsilon(i, j)q_{\varepsilon i}.$$

Putting now

$$f(\tau) = \frac{1}{\varepsilon}a(i, j, x_{0j}, u_0(j) + \tau\varepsilon v(j)), \qquad \tau \in [0, 1],$$

and note that

$$f'(\tau) = \nabla_u a(i, j, x_{0j}, u_0(j) + \tau\varepsilon v(j))v(j),$$

$$\int_0^1 f'(\tau)\mathrm{d}\tau = \frac{1}{\varepsilon}[a(i, j, x_{0j}, u_\varepsilon(j)) - a(i, j, x_{0j}, u_0(j))],$$

we obtain

$$\frac{1}{\varepsilon}[a(i, j, x_{0j}, u_\varepsilon(j)) - a(i, j, x_{0j}, u_0(j))] = a_\varepsilon(i, j)v(j).$$

Continuing similar transformations, we represent (2.30) in the form

$$\begin{aligned}
q_\varepsilon(i+1) &= \Phi_\varepsilon(i+1)q_{\varepsilon,i+1} + \sum_{j=1}^{i} A_\varepsilon(i, j)q_{\varepsilon j} + \sum_{j=1}^{i} B_\varepsilon(i, j)q_{\varepsilon j}\xi(j+1) \\
&+ \sum_{j=0}^{i} a_\varepsilon(i, j)v(j) + \sum_{j=0}^{i} b_\varepsilon(i, j)v(j)\xi(j+1), \quad q_{\varepsilon 0} = 0.
\end{aligned} \tag{2.34}$$

From (2.34) via (2.31), (2.9) it follows that

$$
\begin{aligned}
&(1 - K_0(i+1))|q_\varepsilon(i+1)| \\
&\quad \le \sum_{j=1}^{i} |q_\varepsilon(j)| K_0(j) + \sum_{j=1}^{i} |A_\varepsilon(i,j) q_{\varepsilon j}| + \sum_{j=0}^{i} |a_\varepsilon(i,j) v(j)| \\
&\qquad + \left| \sum_{j=1}^{i} B_\varepsilon(i,j) q_{\varepsilon j} \xi(j+1) \right| + \left| \sum_{j=0}^{i} b_\varepsilon(i,j) v(j) \xi(j+1) \right| .
\end{aligned}
$$

Calculating pth moments of the both parts of this inequality, via Lemma 1.7, we obtain

$$
\begin{aligned}
&(1 - K_0(i+1))^p \mathbf{E}|q_\varepsilon(i+1)|^p \\
&\quad \le 5^{p-1} \left[\mathbf{E} \left(\sum_{j=1}^{i} |q_\varepsilon(j)| K_0(j) \right)^p + \mathbf{E} \left(\sum_{j=1}^{i} |A_\varepsilon(i,j) q_{\varepsilon j}| \right)^p \right. \\
&\qquad + \mathbf{E} \left(\sum_{j=0}^{i} |a_\varepsilon(i,j) v(j)| \right)^p + \mathbf{E} \left| \sum_{j=1}^{i} B_\varepsilon(i,j) q_{\varepsilon j} \xi(j+1) \right|^p \\
&\qquad \left. + \mathbf{E} \left| \sum_{j=0}^{i} b_\varepsilon(i,j) v(j) \xi(j+1) \right|^p \right] .
\end{aligned}
$$

Note also that via Lemma 1.7, (2.13), and $M_p = \max_{j \in Z} \mathbf{E}|\xi(j)|^p$ for $i \le N-1$, we have

$$
\begin{aligned}
\mathbf{E} \left(\sum_{j=1}^{i} |q_\varepsilon(j)| K_0(j) \right)^p &\le N^{p-1} K_0^p \sum_{j=1}^{i} \mathbf{E}|q_\varepsilon(j)|^p, \\
\mathbf{E} \left(\sum_{j=1}^{i} |A_\varepsilon(i,j) q_{\varepsilon j}| \right)^p &\le N^{p-1} \sum_{j=1}^{i} \mathbf{E}|A_\varepsilon(i,j) q_{\varepsilon j}|^p, \\
\mathbf{E} \left(\sum_{j=0}^{i} |a_\varepsilon(i,j) v(j)| \right)^p &\le N^{p-1} \sum_{j=0}^{i} \mathbf{E}|a_\varepsilon(i,j) v(j)|^p, \\
\mathbf{E} \left| \sum_{j=1}^{i} B_\varepsilon(i,j) q_{\varepsilon j} \xi(j+1) \right|^p &\le N^{p-1} \sum_{j=1}^{i} \mathbf{E}|B_\varepsilon(i,j) q_{\varepsilon j}|^p \mathbf{E}|\xi(j+1)|^p
\end{aligned}
$$

$$\leq N^{p-1}M_p \sum_{j=1}^{i} \mathbf{E}|B_\varepsilon(i,j)q_{\varepsilon j}|^p,$$

$$\mathbf{E}\left|\sum_{j=0}^{i} b_\varepsilon(i,j)v(j)\xi(j+1)\right|^p \leq N^{p-1}\sum_{j=0}^{i} \mathbf{E}|b_\varepsilon(i,j)v(j)|^p \mathbf{E}|\xi(j+1)|^p$$
$$\leq N^{p-1}M_p \sum_{j=0}^{i} \mathbf{E}|b_\varepsilon(i,j)v(j)|^p.$$

So, putting $\hat{M}_p = \max\{1, M_p\}$ and using (2.13), we get

$$\mathbf{E}|q_\varepsilon(i+1)|^p \leq \frac{(5N)^{p-1}}{(1-K_0)^p}\left[K_0^p \sum_{j=1}^{i} \mathbf{E}|q_\varepsilon(j)|^p \right.$$
$$+ \hat{M}_p \sum_{j=1}^{i} \mathbf{E}\left(|A_\varepsilon(i,j)q_{\varepsilon j}|^p + |B_\varepsilon(i,j)q_{\varepsilon j}|^p\right)$$
$$\left. + \hat{M}_p \sum_{j=0}^{i} \mathbf{E}\left(|a_\varepsilon(i,j)|^p + |b_\varepsilon(i,j)|^p\right)|v(j)|^p \right].$$

Note that via (2.31)

$$|A_\varepsilon(i,j)q_{\varepsilon j}|^p + |B_\varepsilon(i,j)q_{\varepsilon j}|^p = \left|\int_0^1 \nabla a(i,j,\lambda_{\varepsilon j}^\tau, u_\varepsilon(j))q_{\varepsilon j}\mathrm{d}\tau\right|^p$$
$$+ \left|\int_0^1 \nabla b(i,j,\lambda_{\varepsilon j}^\tau, u_\varepsilon(j))q_{\varepsilon j}\mathrm{d}\tau\right|^p$$
$$\leq \int_0^1 (|\nabla a(i,j,\lambda_{\varepsilon j}^\tau, u_\varepsilon(j))q_{\varepsilon j}|^p$$
$$+ |\nabla b(i,j,\lambda_{\varepsilon j}^\tau, u_\varepsilon(j))q_{\varepsilon j}|^p)\mathrm{d}\tau. \tag{2.35}$$

Via (2.10), (2.13), $q_\varepsilon(0) = 0$, and Lemma 1.7, we obtain

$$|\nabla a(i,j,\lambda_{\varepsilon j}^\tau, u_\varepsilon(j))q_{\varepsilon j}|^p + |\nabla b(i,j,\lambda_{\varepsilon j}^\tau, u_\varepsilon(j))q_{\varepsilon j}|^p$$
$$\leq \left(|\nabla a(i,j,\lambda_{\varepsilon j}^\tau, u_\varepsilon(j))q_{\varepsilon j}|^2 + |\nabla b(i,j,\lambda_{\varepsilon j}^\tau, u_\varepsilon(j))q_{\varepsilon j}|^2\right)^{p/2}$$

$$\leq \left(\sum_{l=1}^{j} |q_\varepsilon(l)|^2 K_1(l) \right)^{p/2}$$

$$\leq K_1^{\frac{p}{2}} N^{\frac{p}{2}-1} \sum_{l=1}^{j} |q_\varepsilon(l)|^p. \tag{2.36}$$

Similarly to (2.35), (2.36) via (2.11) we have

$$|a_\varepsilon(i,j)|^p + |b_\varepsilon(i,j)|^p$$

$$\leq \int_0^1 \left(|\nabla_u a(i,j,x_{0j},u_\varepsilon^\tau(j))|^p + |\nabla_u a(i,j,x_{0j},u_\varepsilon^\tau(j))|^p \right) d\tau$$

$$\leq \int_0^1 \left(|\nabla_u a(i,j,x_{0j},u_\varepsilon^\tau(j))| + |\nabla_u a(i,j,x_{0j},u_\varepsilon^\tau(j))| \right)^p d\tau$$

$$\leq C^p.$$

So, as a result we obtain

$$\mathbf{E}|q_\varepsilon(i+1)|^p \leq \frac{(5N)^{p-1}}{(1-K_0)^p} \left[K_0^p \sum_{j=1}^{i} \mathbf{E}|q_\varepsilon(j)|^p + \hat{M}_p K_1^{\frac{p}{2}} N^{\frac{p}{2}-1} \sum_{l=1}^{i} \sum_{j=l}^{i} \mathbf{E}|q_\varepsilon(l)|^p \right.$$
$$\left. + \hat{M}_p C^p \sum_{j=0}^{i} \mathbf{E}|v(j)|^p \right]$$
$$\leq \frac{(5N)^{p-1}}{(1-K_0)^p} \left[\hat{M}_p C^p \sum_{j=0}^{i} \mathbf{E}|v(j)|^p + C_1 \sum_{j=1}^{i} \mathbf{E}|q_\varepsilon(j)|^p \right],$$

where $C_1 = K_0^p + \hat{M}_p (K_1 N)^{\frac{p}{2}}$. Putting also

$$C_2 = \frac{(5N)^{p-1} C_1}{(1-K_0)^p}, \quad y_\varepsilon(i) = \mathbf{E}|q_\varepsilon(i)|^p, \quad z(i) = \frac{\hat{M}_p C^p}{C_1} \sum_{j=0}^{i} \mathbf{E}|v(j)|^p,$$

rewrite the obtained inequality in the form

$$y_\varepsilon(i+1) \leq C_2 \left[z(i) + \sum_{j=1}^{i} y_\varepsilon(j) \right].$$

Via Lemma 1.2, (1.14), and $\max_{j\in Z}\mathbf{E}|v(j)|^p < \infty$, we have $y_\varepsilon(i+1) \leq C_2(1+C_2)^{N-1}z(i) < \infty$, i.e., q_ε is uniformly p-bounded with respect to $\varepsilon \geq 0$. The proof is completed.

Definition 2.3 The family of random variables $\{\xi_\varepsilon\}_{\varepsilon\geq 0}$ is called uniformly integrable, if $\sup_{\varepsilon\geq 0}\mathbf{E}\left[|\xi_\varepsilon|\chi\{|\xi_\varepsilon| > K\}\right] \to 0$ by $K \to \infty$ (here $\chi\{|\xi_\varepsilon| > K\}$ is the indicator of the set $\{w : |\xi_\varepsilon| > K\}$).

Lemma 2.3 *Let the conditions* (2.9), (2.10), (2.12)–(2.15) *hold and the auxiliary process* $v(j)$ *in* (2.29) *satisfies the condition*

$$\max_{i\in Z}\mathbf{E}|v(i)|^4 < \infty. \tag{2.37}$$

Then the process $|q_\varepsilon(i)|^2$ *is uniformly integrable with respect to* $\varepsilon \geq 0$ *and* $\lim_{\varepsilon\to 0}\|q_\varepsilon - q_0\|_i = 0$, $i = 1, \ldots, N$.

Proof Via (2.37) from Lemma 2.2 it follows that the process $\mathbf{E}|q_\varepsilon(i)|^4$ is uniformly bounded with respect to $\varepsilon \geq 0$. From this and [67, 184], it follows that the process $|q_\varepsilon(i)|^2$ is uniformly integrable with respect to $\varepsilon \geq 0$. Let us prove the second statement of the lemma. From (2.34), it follows that

$$\begin{aligned}
q_\varepsilon(i+1) - q_0(i+1) &= \Phi_\varepsilon(i+1)q_{\varepsilon,i+1} - \Phi_0(i+1)q_{0,i+1} \\
&\quad + \sum_{j=1}^{i}\left[A_\varepsilon(i,j)q_{\varepsilon j} - A_0(i,j)q_{0j}\right] \\
&\quad + \sum_{j=1}^{i}\left[B_\varepsilon(i,j)q_{\varepsilon j} - B_0(i,j)q_{0j}\right]\xi(j+1) \\
&\quad + \sum_{j=0}^{i}\left[a_\varepsilon(i,j) - a_0(i,j)\right]v(j) \\
&\quad + \sum_{j=0}^{i}\left[b_\varepsilon(i,j) - b_0(i,j)\right]v(j)\xi(j+1).
\end{aligned}$$

Rewrite this representation in the form

$$\begin{aligned}
q_\varepsilon(i+1) - q_0(i+1) &= \Phi_0(i+1)(q_{\varepsilon,i+1} - q_{0,i+1}) + (\Phi_\varepsilon(i+1) - \Phi_0(i+1))q_{\varepsilon,i+1} \\
&\quad + \sum_{j=1}^{i}[A_0(i,j)(q_{\varepsilon j} - q_{0j}) + (A_\varepsilon(i,j) - A_0(i,j))q_{\varepsilon j}] \\
&\quad + \sum_{j=1}^{i}[B_0(i,j)(q_{\varepsilon j} - q_{0j}) + (B_\varepsilon(i,j) - B_0(i,j))q_{\varepsilon j}]\xi(j+1)
\end{aligned}$$

$$+\sum_{j=0}^{i}[a_\varepsilon(i,j)-a_0(i,j)]v(j)$$

$$+\sum_{j=0}^{i}[b_\varepsilon(i,j)-b_0(i,j)]v(j)\xi(j+1).$$

From this via (2.9) it follows that

$$\begin{aligned}
&(1-K_0(i+1))|q_\varepsilon(i+1)-q_0(i+1)| \\
&\quad\le \sum_{j=1}^{i}|q_\varepsilon(j)-q_0(j)|K_0(j)+|(\Phi_\varepsilon(i+1)-\Phi_0(i+1))q_{\varepsilon,i+1}| \\
&\quad+\sum_{j=1}^{i}|A_0(i,j)(q_{\varepsilon j}-q_{0j})|+\sum_{j=0}^{i}|(A_\varepsilon(i,j)-A_0(i,j))q_{\varepsilon j}| \\
&\quad+\left|\sum_{j=1}^{i}B_0(i,j)(q_{\varepsilon j}-q_{0j})\xi(j+1)\right|+\left|\sum_{j=1}^{i}(B_\varepsilon(i,j)-B_0(i,j))q_{\varepsilon j}\xi(j+1)\right| \\
&\quad+\sum_{j=0}^{i}|(a_\varepsilon(i,j)-a_0(i,j))v(j)|+\left|\sum_{j=0}^{i}[b_\varepsilon(i,j)-b_0(i,j)]v(j)\xi(j+1)\right|.
\end{aligned}$$

Squaring and calculating the expectation, from this via Lemma 1.7 and (2.13) for $i\le N-1$, we obtain

$$\begin{aligned}
&(1-K_0)^2\mathbf{E}|q_\varepsilon(i+1)-q_0(i+1)|^2 \\
&\quad\le 8\Bigg[NK_0^2\sum_{j=1}^{i}\mathbf{E}|q_\varepsilon(j)-q_0(j)|^2+\alpha_1(\varepsilon,i) \\
&\qquad+N\sum_{j=1}^{i}\mathbf{E}|A_0(i,j)(q_{\varepsilon j}-q_{0j})|^2+N\alpha_2(\varepsilon,i) \\
&\qquad+\sum_{j=1}^{i}\mathbf{E}|B_0(i,j)(q_{\varepsilon j}-q_{0j})|^2+N\alpha_3(\varepsilon,i)\Bigg],
\end{aligned}\qquad (2.38)$$

where

$$\alpha_1(\varepsilon,i)=\mathbf{E}|(\Phi_\varepsilon(i+1)-\Phi_0(i+1))q_{\varepsilon,i+1}|^2,$$

$$\alpha_2(\varepsilon,i)=\sum_{j=1}^{i}\mathbf{E}|(A_\varepsilon(i,j)-A_0(i,j))q_{\varepsilon j}|^2+\sum_{j=1}^{i}\mathbf{E}|(B_\varepsilon(i,j)-B_0(i,j))q_{\varepsilon j}|^2,$$

$$\alpha_3(\varepsilon, i) = \sum_{j=0}^{i} \mathbf{E}|(a_\varepsilon(i, j) - a_0(i, j))v(j)|^2 + \sum_{j=0}^{i} \mathbf{E}|(b_\varepsilon(i, j) - b_0(i, j))v(j)|^2. \tag{2.39}$$

So, similarly to (2.35), (2.36) with $p = 2$ from (2.38) it follows that

$$\begin{aligned}
&(1 - K_0)^2\mathbf{E}|q_\varepsilon(i + 1) - q_0(i + 1)|^2 \\
&\quad\le 8N\left[\sum_{k=1}^{3} \alpha_k(\varepsilon, i) + K_0^2 \sum_{j=1}^{i} \mathbf{E}|q_\varepsilon(j) - q_0(j)|^2 \right.\\
&\qquad\left. + \sum_{j=1}^{i} \mathbf{E}\left(|A_0(i, j)(q_{\varepsilon j} - q_{0j})|^2 + |B_0(i, j)(q_{\varepsilon j} - q_{0j})|^2\right)\right] \\
&\quad\le 8N\left[\sum_{k=1}^{3} \alpha_k(\varepsilon, i) + (K_0^2 + K_1 N) \sum_{j=1}^{i} \mathbf{E}|q_\varepsilon(j) - q_0(j)|^2\right]
\end{aligned}$$

or

$$y_\varepsilon(i + 1) \le C_2\left[z_\varepsilon(i) + \sum_{j=1}^{i} y_\varepsilon(j)\right], \tag{2.40}$$

where

$$C_2 = \frac{8NC_1}{(1 - K_0)^2}, \qquad z_\varepsilon(i) = \frac{1}{C_1}\sum_{k=1}^{3} \alpha_k(\varepsilon, i),$$

$$C_1 = K_0^2 + K_1 N, \qquad y_\varepsilon(i) = \mathbf{E}|q_\varepsilon(i) - q_0(i)|^2.$$

Via Lemma 1.2 we have $y_\varepsilon(i + 1) \le C_2(1 + C_2)^{N-1}z_\varepsilon(i)$. So, it is enough to show that $\lim_{\varepsilon\to 0} z_\varepsilon(i) = 0$, $i \in Z$.

Let us show from the beginning that $\lim_{\varepsilon\to 0} \alpha_1(\varepsilon, i) = 0$, $i \in Z$. Represent $q_\varepsilon(i)$ in the form

$$q_\varepsilon(i) = q_\varepsilon(i)\chi_\varepsilon^K(i) + q_\varepsilon(i)(1 - \chi_\varepsilon^K(i)), \tag{2.41}$$

where $\chi_\varepsilon^K(i)$ is the indicator of the set $\{\omega : |q_\varepsilon(i)| > K\}$.

Then via Lemma 1.7

$$\begin{aligned}
\alpha_1(\varepsilon, i - 1) &= \mathbf{E}\,|(\Phi_\varepsilon(i) - \Phi_0(i))q_{\varepsilon i}|^2 \\
&= \mathbf{E}\left|\Phi_\varepsilon(i)\left(q_\varepsilon\chi_\varepsilon^K\right)_i - \Phi_0(i)\left(q_\varepsilon\chi_\varepsilon^K\right)_i\right.
\end{aligned}$$

$$
\begin{aligned}
&+ (\Phi_\varepsilon(i) - \Phi_0(i)) \left(q_\varepsilon \left(1 - \chi_\varepsilon^K\right)\right)_i \Big|^2 \\
&\le 3 \left[\mathbf{E} \left| \Phi_\varepsilon(i) \left(q_\varepsilon \chi_\varepsilon^K\right)_i \right|^2 + \mathbf{E} \left| \Phi_0(i) \left(q_\varepsilon \chi_\varepsilon^K\right)_i \right|^2 \right. \\
&\left. + \mathbf{E} \left| (\Phi_\varepsilon(i) - \Phi_0(i)) \left(q_\varepsilon \left(1 - \chi_\varepsilon^K\right)\right)_i \right|^2 \right].
\end{aligned}
$$

From this via (2.31), (2.9), (2.12) for $i \le N$ we obtain that

$$
\begin{aligned}
\alpha_1(\varepsilon, i-1) &\le 3 \left[2(N+1) \sum_{j=1}^{i} \mathbf{E} |q_\varepsilon(j) \chi_\varepsilon^K(j)|^2 K_0^2(j) \right. \\
&\left. + \sum_{j=1}^{i} \mathbf{E} |\varepsilon q_\varepsilon(j)|^2 |q_\varepsilon(j)(1 - \chi_\varepsilon^K(j))|^2 K_0(j) \right] \\
&\le 3K_0 \left[2K_0(N+1) \sum_{j=1}^{i} \mathbf{E} |q_\varepsilon(j) \chi_\varepsilon^K(j)|^2 \right. \\
&\left. + \varepsilon^2 \sum_{j=1}^{i} \mathbf{E} |q_\varepsilon(j)|^2 |q_\varepsilon(j) \left(1 - \chi_\varepsilon^K(j)\right)|^2 \right] \\
&\le 3K_0(N+1) \left[2K_0(N+1) \|q_\varepsilon \chi_\varepsilon^K\|_N^2 + \varepsilon^2 K^4 \right].
\end{aligned}
$$

Since $|q_\varepsilon(i)|^2$ is uniformly integrable with respect to $\varepsilon \ge 0$ then $\lim\limits_{K\to\infty} \|q_\varepsilon \chi_\varepsilon^K\|_N = 0$. So, for arbitrary $\delta > 0$ there exists $K > 0$ such that $\|q_\varepsilon \chi_\varepsilon^K\|_N^2 < \delta(12K_0^2(N+1)^2)^{-1}$. Fixing this K, let us choose $\varepsilon > 0$ such that $\varepsilon^2 K^4 < \delta(6K_0(N+1))^{-1}$. Therefore, for arbitrary $\delta > 0$ there exists $\varepsilon > 0$ such that $\alpha_1(\varepsilon, i-1) < \delta$, i.e., $\lim\limits_{\varepsilon\to 0} \alpha_1(\varepsilon, i-1) = 0$.

Note now that via (2.39), (2.41) we have

$$
\begin{aligned}
\alpha_2(\varepsilon, i) = \sum_{j=1}^{i} \mathbf{E} &\left[|A_\varepsilon(i,j) \left(q_\varepsilon \chi_\varepsilon^K\right)_j - A_0(i,j) \left(q_\varepsilon \chi_\varepsilon^K\right)_j \right. \\
&+ (A_\varepsilon(i,j) - A_0(i,j))(q_\varepsilon(1 - \chi_\varepsilon^K))_j|^2 \\
&+ |B_\varepsilon(i,j) \left(q_\varepsilon \chi_\varepsilon^K\right)_j - B_0(i,j) \left(q_\varepsilon \chi_\varepsilon^K\right)_j \\
&\left. + (B_\varepsilon(i,j) - B_0(i,j)) \left(q_\varepsilon(1 - \chi_\varepsilon^K)\right)_j |^2 \right]
\end{aligned}
$$

$$\le 3\sum_{j=1}^{i}\mathbf{E}\Big[|A_\varepsilon(i,j)\left(q_\varepsilon\chi_\varepsilon^K\right)_j|^2+|A_0(i,j)\left(q_\varepsilon\chi_\varepsilon^K\right)_j|^2$$
$$+|(A_\varepsilon(i,j)-A_0(i,j))\left(q_\varepsilon(1-\chi_\varepsilon^K)\right)_j|^2$$
$$+|B_\varepsilon(i,j)\left(q_\varepsilon\chi_\varepsilon^K\right)_j|^2+|B_0(i,j)\left(q_\varepsilon\chi_\varepsilon^K\right)_j|^2$$
$$+|(B_\varepsilon(i,j)-B_0(i,j))\left(q_\varepsilon(1-\chi_\varepsilon^K)\right)_j|^2\Big]$$

or in another form

$$\alpha_2(\varepsilon,i)\le 3\sum_{j=1}^{i}\mathbf{E}\Big[|A_\varepsilon(i,j)\left(q_\varepsilon\chi_\varepsilon^K\right)_j|^2+|B_\varepsilon(i,j)\left(q_\varepsilon\chi_\varepsilon^K\right)_j|^2$$
$$+|A_0(i,j)\left(q_\varepsilon\chi_\varepsilon^K\right)|^2+|B_0(i,j)\left(q_\varepsilon\chi_\varepsilon^K\right)_j|^2$$
$$+|(A_\varepsilon(i,j)-A_0(i,j))(q_\varepsilon(1-\chi_\varepsilon^K))_j|^2$$
$$+|(B_\varepsilon(i,j)-B_0(i,j))(q_\varepsilon(1-\chi_\varepsilon^K))_j|^2\Big].$$

From this and (2.31), (2.10), (2.14) we obtain

$$\alpha_2(\varepsilon,i)\le 3\sum_{j=1}^{i}\int_0^1\mathbf{E}\Big[|\nabla a(i,j,\lambda_{\varepsilon j}^{\tau},u_\varepsilon(j))\left(q_\varepsilon\chi_\varepsilon^K\right)_j|^2$$
$$+|\nabla b(i,j,\lambda_{\varepsilon j}^{\tau},u_\varepsilon(j))\left(q_\varepsilon\chi_\varepsilon^K\right)_j|^2\Big]d\tau$$
$$+3\sum_{j=1}^{i}\mathbf{E}\Big[|\nabla a(i,j,x_{0j},u_0(j))\left(q_\varepsilon\chi_\varepsilon^K\right)_j|^2$$
$$+|\nabla b(i,j,x_{0j},u_0(j))\left(q_\varepsilon\chi_\varepsilon^K\right)_j|^2\Big]$$
$$+3\sum_{j=1}^{i}\int_0^1\mathbf{E}\Big[|(\nabla a(i,j,\lambda_{\varepsilon j}^{\tau},u_\varepsilon(j))-\nabla a(i,j,x_{0j},u_0(j)))$$
$$\times\left(q_\varepsilon(1-\chi_\varepsilon^K)\right)_j|^2+|(\nabla b(i,j,\lambda_{\varepsilon j}^{\tau},u_\varepsilon(j))$$
$$-\nabla b(i,j,x_{0j},u_0(j)))\left(q_\varepsilon(1-\chi_\varepsilon^K)\right)_j|^2\Big]d\tau$$

$$
\begin{aligned}
&\le 3\sum_{j=1}^{i}\sum_{l=1}^{j}\Big[2\mathbf{E}|q_\varepsilon(l)\chi_\varepsilon^K(l)|^2 K_1(l) + \varepsilon^2\mathbf{E}\left(|q_\varepsilon(l)|^2 + |v(j)|^2\right) \\
&\qquad\qquad \times |q_\varepsilon(l)(1-\chi_\varepsilon^K(l))|^2 K_1(l)\Big] \\
&\le \frac{3}{2}K_1(N+1)(N+2)\left[2\|q_\varepsilon\chi_\varepsilon^K\|_N^2 + \varepsilon^2 K^2\left(K^2 + \|v\|^2\right)\right].
\end{aligned}
$$

Since $|q_\varepsilon(i)|^2$ is uniformly integrable with respect to $\varepsilon \ge 0$ then $\lim\limits_{K\to\infty}\|q_\varepsilon\chi_\varepsilon^K\|_N = 0$. So, for arbitrary $\delta > 0$ there exists $K > 0$ such that $\|q_\varepsilon\chi_\varepsilon^K\|_N^2 < \delta(6K_1(N+1)(N+2))^{-1}$. Fixing this K, let us choose $\varepsilon > 0$ such that $\varepsilon^2 K^2(K_2 + \|v\|^2) < \delta(3K_1(N+1)(N+2))^{-1}$. Therefore, for arbitrary $\delta > 0$ there exists $\varepsilon > 0$ such that $\alpha_2(\varepsilon, i) < \delta$, i.e., $\lim\limits_{\varepsilon\to 0}\alpha_2(\varepsilon, i) = 0$.

Similarly, via (2.31) and (2.15) one can get that

$$
\begin{aligned}
\alpha_3(\varepsilon, i) &\le \sum_{j=0}^{i}\mathbf{E}\left(|a_\varepsilon(i,j) - a_0(i,j)|^2 + |b_\varepsilon(i,j) - b_0(i,j)|^2\right)|v(j)|^2 \\
&\le \sum_{j=0}^{i}\mathbf{E}\Bigg[\int_0^1 |\nabla_u a(i,j,x_{0j},u_\varepsilon^\tau(j)) - \nabla_u a(i,j,x_{0j},u_0(j))|^2 d\tau \\
&\qquad + \int_0^1 |\nabla_u b(i,j,x_{0j},u_\varepsilon^\tau(j)) - \nabla_u b(i,j,x_{0j},u_0(j))|^2 d\tau\Bigg]|v(j)|^2 \\
&\le \sum_{j=0}^{i}\mathbf{E}\int_0^1 \big[|\nabla_u a(i,j,x_{0j},u_\varepsilon^\tau(j)) - \nabla_u a(i,j,x_{0j},u_0(j))| \\
&\qquad + |\nabla_u b(i,j,x_{0j},u_\varepsilon^\tau(j)) - \nabla_u b(i,j,x_{0j},u_0(j))|\big]^2 d\tau|v(j)|^2 \\
&\le \varepsilon^2 C^2\sum_{j=0}^{i}\mathbf{E}|v(j)|^4.
\end{aligned}
$$

So, via (2.37) $\lim\limits_{\varepsilon\to 0}\alpha_3(\varepsilon, i) = 0$.

As a result, we have $\lim\limits_{\varepsilon\to 0} z_\varepsilon(i) = 0$, $i \in Z$, and, therefore, $\lim\limits_{\varepsilon\to 0}\|q_\varepsilon - q_0\|_i^2 = 0$. The proof is completed.

2.1.3 Main Result

In this section, the bounded limit (1.9), (1.10) for the optimal control problem (2.3), (2.4) will be calculated. So, the necessary condition $J_0(u_0) \geq 0$ for optimality of the control u_0 will be obtained.

Theorem 2.1 *Let the conditions* (2.5)–(2.15), (2.18)–(2.24), (2.37) *hold. Then the limit* (1.9), (1.10) *for the optimal control problem* (2.3), (2.4) *there exists, equals*

$$J_0(u_0) = \mathbf{E}\left[\langle F_0(N), q_{0N}\rangle + \sum_{j=0}^{N-1} \left(\langle G_0(j), q_{0j}\rangle + v'(j)g_0(j)\right)\right] \tag{2.42}$$

and it is bounded.

Here, $x_0(j)$ *is the solution of the equation* (2.3) *under the control* $u_0(j)$, $q_0(j)$ *is the solution of the linear stochastic difference equation*

$$\begin{aligned} q_0(i+1) &= \eta_0(i+1) + \Phi_0(i+1)q_{0,i+1} \\ &\quad + \sum_{j=1}^{i} A_0(i,j)q_{0j} + \sum_{j=1}^{i} B_0(i,j)q_{0j}\xi(j+1), \\ q_{00} &= 0, \end{aligned} \tag{2.43}$$

where

$$\eta_0(i+1) = \sum_{j=0}^{i} a_0(i,j)v(j) + \sum_{j=0}^{i} b_0(i,j)v(j)\xi(j+1), \tag{2.44}$$

$$\begin{aligned} F_0(N) &= \nabla F(x_{0N}), & \Phi_0(i) &= \nabla\Phi(i, x_{0i}), \\ G_0(j) &= \nabla G(j, x_{0j}, u_0(j)), & g_0(j) &= \nabla_u G(j, x_{0j}, u_0(j)), \\ A_0(i,j) &= \nabla a(i,j,x_{0j},u_0(j)), & a_0(i,j) &= \nabla_u a(i,j,x_{0j},u_0(j)), \\ B_0(i,j) &= \nabla b(i,j,x_{0j},u_0(j)), & b_0(i,j) &= \nabla_u b(i,j,x_{0j},u_0(j)). \end{aligned} \tag{2.45}$$

Proof From (2.4), it follows that

$$\begin{aligned} J_\varepsilon(u_0) &= \frac{1}{\varepsilon}\left[J(u_\varepsilon) - J(u_0)\right] \\ &= \mathbf{E}\Bigg[\frac{1}{\varepsilon}(F(x_{\varepsilon N}) - F(x_{0N})) + \frac{1}{\varepsilon}\sum_{j=0}^{N-1}(G(j, x_{\varepsilon j}, u_\varepsilon(j)) - G(j, x_{0j}, u_\varepsilon(j))) \\ &\quad + \frac{1}{\varepsilon}\sum_{j=0}^{N-1}(G(j, x_{0j}, u_\varepsilon(j)) - G(j, x_{0j}, u_0(j)))\Bigg]. \end{aligned}$$

To transform the expression in square brackets, put

$$F_\varepsilon(N) = \int_0^1 \nabla F(\lambda_{\varepsilon N}^\tau)\mathrm{d}\tau,$$

$$G_\varepsilon(j) = \int_0^1 \nabla G(j, \lambda_{\varepsilon j}^\tau, u_\varepsilon(j))\mathrm{d}\tau,$$

$$g_\varepsilon(j) = \int_0^1 \nabla_u G(j, x_{0j}, u_\varepsilon^\tau(j))\mathrm{d}\tau, \tag{2.46}$$

and consider the function

$$f(\tau) = \frac{1}{\varepsilon} F(x_{0N} + \varepsilon\tau q_{\varepsilon N}), \qquad \tau \in [0, 1].$$

Since $f'(\tau) = \langle \nabla F(\lambda_{\varepsilon N}^\tau), q_{\varepsilon N}\rangle$, where λ_ε^τ is defined in (2.31), then via (2.46)

$$\frac{1}{\varepsilon}[F(x_{\varepsilon N}) - F(x_{0N})] = \langle F_\varepsilon(N), q_{\varepsilon N}\rangle .$$

Similarly, we have

$$\frac{1}{\varepsilon}\left[G(j, x_{\varepsilon j}, u_\varepsilon(j)) - G(j, x_{0j}, u_\varepsilon(j))\right] = \langle G_\varepsilon(j)), q_{\varepsilon j}\rangle .$$

To transform the difference

$$\frac{1}{\varepsilon}\left[G(j, x_{0j}, u_\varepsilon(j)) - G(j, x_{0j}, u_0(j))\right],$$

consider the function

$$f(\tau) - \frac{1}{\varepsilon} G(j, x_{0j}, u_0(j) + \varepsilon\tau v(j)), \qquad \tau \in [0, 1],$$

for which $f'(\tau) = v'(j)\nabla_u G(j, x_{0j}, u_\varepsilon^\tau(j))$. So, via (2.46) we obtain

$$\frac{1}{\varepsilon}\left[G(j, x_{0j}, u_\varepsilon(j)) - G(j, x_{0j}, u_0(j))\right] = v'(j) g_\varepsilon(j).$$

Thus, $J_\varepsilon(u_0)$ takes the representation

$$J_\varepsilon(u_0) = \mathbf{E}\left[\langle F_\varepsilon(N), q_{\varepsilon N}\rangle + \sum_{j=0}^{N-1}\left(\langle G_\varepsilon(j), q_{\varepsilon j}\rangle + v'(j) g_\varepsilon(j)\right)\right].$$

Rewrite $J_\varepsilon(u_0)$ in the form

$$J_\varepsilon(u_0) = \mathbf{E}\left[\langle F_0(N), q_{\varepsilon N}\rangle + \sum_{j=0}^{N-1}\big\langle G_0(j), q_{\varepsilon j}\big\rangle + v'(j)g_0(j)\right] + \sum_{i=1}^{3}\beta_i(\varepsilon),$$

where $F_0(N)$, $G_0(j)$, $g_0(j)$ are defined in (2.45) and

$$\beta_1(\varepsilon) = \mathbf{E}\,\langle F_\varepsilon(N) - F_0(N), q_{\varepsilon N}\rangle\,,$$

$$\beta_2(\varepsilon) = \mathbf{E}\sum_{j=1}^{N-1}\big\langle G_\varepsilon(j) - G_0(j), q_{\varepsilon j}\big\rangle\,,$$

$$\beta_3(\varepsilon) = \mathbf{E}\sum_{j=0}^{N-1} v'(j)(g_\varepsilon(j) - g_0(j)).$$

Let us show that $\lim_{\varepsilon\to 0}\beta_i(\varepsilon) = 0, i = 1, 2, 3$. Via (2.46), (2.20), (2.31), (2.13) we have

$$\begin{aligned}
|\beta_1(\varepsilon)| &\le \mathbf{E}\int_0^1 |\langle\nabla F(\lambda_{\varepsilon N}^\tau) - \nabla F(x_{0N}), q_{\varepsilon N}\rangle|d\tau \\
&\le \varepsilon\sum_{j=1}^{N}\mathbf{E}|q_\varepsilon(j)|^2 K_1(j) \\
&\le \varepsilon K_1 N\|q_\varepsilon\|_N^2.
\end{aligned}$$

From this via Lemma 2.2, it follows that $\lim_{\varepsilon\to 0}\beta_1(\varepsilon) = 0$.

Similarly, using (2.46), (2.21), (2.31), (2.29), (2.13), (2.23) we get

$$\begin{aligned}
|\beta_2(\varepsilon)| &\le \mathbf{E}\sum_{j=1}^{N-1}\int_0^1 |\langle\nabla G(j, \lambda_{\varepsilon j}^\tau, u_\varepsilon(j)) - \nabla G(j, x_{0j}, u_0(j)), q_{\varepsilon j}\rangle|d\tau \\
&\le \varepsilon\mathbf{E}\sum_{j=1}^{N-1}\sum_{l=1}^{j}(|q_\varepsilon(l)| + |v(j)|)|q_\varepsilon(l)|K_1(l) \\
&\le \frac{1}{2}\varepsilon K_1(N-1)^2\left(3\|q_\varepsilon\|_N^2 + \|v\|_N^2\right),
\end{aligned}$$

and

$$\begin{aligned}
|\beta_3(\varepsilon)| &\le \mathbf{E}\sum_{j=0}^{N-1}|v(j)|\int_0^1 |\nabla_u G(j, x_{0j}, u_\varepsilon^\tau(j)) - \nabla_u G(j, x_{0j}, u_0(j))|d\tau \\
&\le \varepsilon CN\|v\|_N^2.
\end{aligned}$$

From this and Lemmas 2.2 and 2.3, we obtain $\lim_{\varepsilon\to 0} |J_\varepsilon(u_0) - J_0(u_0)| = 0$. From the conditions (2.18), (2.19), (2.22), it follows also that $J_0(u_0) < \infty$. The proof is completed.

2.2 A Linear-Quadratic Problem

Here, the necessary condition for control optimality obtained above is used to construct, in an explicit form, the synthesis of optimal control for a linear-quadratic problem.

2.2.1 *Synthesis of Optimal Control*

Consider the optimal control problem for the linear equation

$$x(i+1) = \eta(i+1) + \sum_{j=0}^{i} a(i,j)x(j) + \sum_{j=0}^{i} b(i,j)u(j), \quad i = 0, \ldots, N-1, \tag{2.47}$$

with the initial condition $x(0) = \varphi_0(0)$ and the performance functional

$$J(u) = \mathbf{E}\left[x'(N)Fx(N) + \sum_{j=0}^{N-1} u'(j)G(j)u(j)\right]. \tag{2.48}$$

Here $\eta \in H$, $x(i) \in \mathbf{R}^n$, $u(i) \in \mathbf{R}^m$, positive semidefinite matrix F, positive definite matrix $G(j)$, and matrices $a(i,j)$, $b(i,j)$, $i,j \in Z$, are nonrandom matrices of appropriate dimensions.

The optimal control problem (2.47), (2.48) is a particular case of the control problem (2.3), (2.4) with

$$\begin{gathered}
\Phi(i+1, x_{i+1}) = 0, \qquad b(i,j,x_j,u(j)) = 0, \\
a(i,j,x_j,u(j)) = a(i,j)x(j) + b(i,j)u(j), \\
F(x_N) = x'(N)Fx(N), \qquad G(j,x_j,u(j)) = u'(j)G(j)u(j).
\end{gathered} \tag{2.49}$$

Let us calculate the coefficients of the Eqs. (2.43)–(2.45) for the optimal control problem (2.47), (2.48). Via (2.49), (2.45) we have

$$\begin{gathered}
\Phi_0(i+1) = 0, \qquad B_0(i,j) = 0, \qquad b_0(i,j) = 0, \\
A_0(i,j)q_{0j} = a(i,j)q_0(j), \qquad a_0(i,j)v(j) = b(i,j)v(j).
\end{gathered}$$

From this and (2.43), (2.44) it follows that

$$q_0(i+1)=\sum_{j=1}^{i}a(i,j)q_0(j)+\sum_{j=0}^{i}b(i,j)v(j),$$
$$q_0(0)=0. \tag{2.50}$$

Besides, via (2.49), (2.45) we obtain

$$\langle F_0(N),q_{0N}\rangle=2x_0'(N)Fq_0(N),$$
$$\langle G_0(j),q_{0j}\rangle=0,\qquad v'(j)g_0(j)=2v'(j)G(j)u_0(j).$$

Via Theorem 2.1 from this and (2.42), it follows that the limit (1.9), (1.10) for the optimal control problem (2.47), (2.48) there exists and equals

$$J_0(u_0)=2\mathbf{E}\left[x_0'(N)Fq_0(N)+\sum_{j=0}^{N-1}u_0'(j)G(j)v(j)\right]. \tag{2.51}$$

Here, x_0 is a solution of the equation (2.47) under the control u_0 and $q_0(i)$ is a solution of the equation (2.50).

Note that for the linear-quadratic problem (2.47), (2.48) $q_\varepsilon\equiv q_0$. So, Lemma 2.3 became a trivial one and the condition of boundedness of the process $\mathbf{E}|q_0(i)|^2$ is a sufficient condition for existence of (2.51).

Definition 2.3 For arbitrary function $f(k)$, $0\le j\le k\le i$, the functional $\psi(i,j,f(\cdot))$ is defined by the following way

$$\psi(i,j,f(\cdot))=f(i)+\sum_{k=j+1}^{i}R(i,k)f(k-1), \tag{2.52}$$

where $R(i,k)$ is the resolvent of the kernel $a(i,k)$ from the Eq. (2.47).

Lemma 2.4 *The necessary condition* $J_0(u_0)\ge 0$ *for optimality of the control* u_0 *of the problem* (2.47), (2.48) *has a unique solution*

$$u_0(j)=-G^{-1}(j)\psi'(N-1,j,b(\cdot,j))F\mathbf{E}_jx_0(N),\quad j=0,\ldots,N-1. \tag{2.53}$$

Proof Using the resolvent $R(i,k)$ of the kernel $a(i,k)$ and the representations (1.17), (2.52), rewrite the solution of the equation (2.50) in the form

$$\begin{aligned} q_0(i+1) &= \sum_{j=0}^{i} b(i,j)v(j) + \sum_{k=1}^{i} R(i,k) \sum_{j=0}^{k-1} b(k-1,j)v(j) \\ &= \sum_{j=0}^{i} \left[b(i,j) + \sum_{k=j+1}^{i} R(i,k)b(k-1,j) \right] v(j) \\ &= \sum_{j=0}^{i} \psi(i,j,b(\cdot,j))v(j). \end{aligned} \tag{2.54}$$

Substituting (2.54) into (2.51), we obtain

$$J_0(u_0) = 2\mathbf{E} \sum_{j=0}^{N-1} \left[\mathbf{E}_j x_0'(N) F \psi(N-1,j,b(\cdot,j)) + u_0'(j)G(j) \right] v(j). \tag{2.55}$$

The inequality $J_0(u_0) \geq 0$ holds for any $v \in \mathbf{U}$ if and only if the expression in the square brackets in (2.55) equals zero, that is equivalent to (2.53). The proof is completed.

Theorem 2.2 *The optimal control of the control problem* (2.47), (2.48) *is represented in the form*

$$\begin{aligned} u_0(0) &= p(0)\left[\psi(N-1,0,\mathbf{E}_0\eta(\cdot+1)) + R(N-1,0)\varphi_0(0)\right], \\ u_0(j+1) &= \alpha(j+1) + p(j+1)\psi(N-1,j,I)x_0(j+1) \\ &\quad + \sum_{k=0}^{j} \gamma(j,k)x_0(k) + Q(j,0)p(0)\left[\psi(N-1,0,\mathbf{E}_0\eta(\cdot+1))\right. \\ &\quad \left. + R(N-1,0)\varphi_0(0)\right], \quad j = 0,1,\ldots,N-2. \end{aligned} \tag{2.56}$$

Here, I is the identical matrix,

$$\begin{aligned} p(j) &= -G^{-1}(j)\psi'(N-1,j,b(\cdot,j))F \\ &\quad \times \left[I + \sum_{k=j}^{N-1} \psi(N-1,k,b(\cdot,k))G^{-1}(k)\psi'(N-1,k,b(\cdot,k))F \right]^{-1}, \\ & j = 0,1,\ldots,N-1, \end{aligned} \tag{2.57}$$

$$\begin{aligned} \alpha(j+1) &= p(j+1)\psi(N-1,j,\beta(\cdot,j+1)) \\ &\quad + \sum_{k=1}^{j} Q(j,k)p(k)\psi(N-1,k-1,\beta(\cdot,k)), \quad j \geq 0, \end{aligned} \tag{2.58}$$

$$\beta(i,j) = \mathbf{E}_j\eta(i+1) - \eta(j), \qquad i \geq j > 0, \tag{2.59}$$

$$\gamma(j,k)=\begin{cases} p(j+1)\psi(N-1,j,a_j(\cdot,k))+Q(j,k)p(k)\psi(N-1,k-1,I) \\ \quad +\sum\limits_{l=k+1}^{j} Q(j,l)p(l)\psi(N-1,l-1,a_{l-1}(\cdot,k)), \qquad j\ge k\ge 1, \\ p(j+1)\psi(N-1,j,a_j(\cdot,0)) \\ \quad +\sum\limits_{l=1}^{j} Q(j,l)p(l)\psi(N-1,l-1,a_{l-1}(\cdot,0)), \qquad j\ge k=0, \end{cases} \tag{2.60}$$

$Q(j,k)$ is the resolvent of the kernel $p(j+1)\psi(N-1,j,b_j(\cdot,k))$,

$$\begin{aligned} a_j(i,k) &= a(i,k)-a(j,k), \\ b_j(i,k) &= b(i,k)-b(j,k), \\ 0 \le k &\le j \le i \le N-1. \end{aligned} \tag{2.61}$$

Proof From (2.53), it follows that to get the representation of the optimal control $u_0(j)$ in the form (2.56), it is enough to calculate the conditional mathematical expectation $\mathbf{E}_j x_0(N)$.

To get $u_0(0)$ from (2.47), we have

$$\mathbf{E}_0 x(i+1)=\zeta(i+1)+\sum_{j=0}^{i} a(i,j)\mathbf{E}_0 x(j), \quad i=0,\dots,N-1, \tag{2.62}$$

where

$$\zeta(i+1)=\mathbf{E}_0\eta(i+1)+\sum_{j=0}^{i} b(i,j)\mathbf{E}_0 u(j), \quad i\ge 0, \quad \zeta(0)=x(0)=\varphi_0(0). \tag{2.63}$$

Using the resolvent $R(i,j)$ of the kernel $a(i,j)$, via (1.16), (1.17) from (2.62), we obtain

$$\mathbf{E}_0 x(i+1)=\zeta(i+1)+\sum_{k=1}^{i} R(i,k)\zeta(k)+R(i,0)\varphi_0(0), \quad i=0,\dots,N-1. \tag{2.64}$$

Substituting (2.63) into (2.64), we get

$$
\begin{aligned}
\mathbf{E}_0 x(i+1) &= \mathbf{E}_0 \eta(i+1) + \sum_{j=0}^{i} b(i,j)\mathbf{E}_0 u(j) + R(i,0)\varphi_0(0) \\
&\quad + \sum_{k=1}^{i} R(i,k)\left[\mathbf{E}_0 \eta(k) + \sum_{j=0}^{k-1} b(k-1,j)\mathbf{E}_0 u(j)\right] \\
&= \mathbf{E}_0 \eta(i+1) + \sum_{k=1}^{i} R(i,k)\mathbf{E}_0 \eta(k) + R(i,0)\varphi_0(0) \\
&\quad + \sum_{j=0}^{i}\left[b(i,j) + \sum_{k=j+1}^{i} R(i,k)b(k-1,j)\right]\mathbf{E}_0 u(j).
\end{aligned}
$$

From this via (2.52) for $f(i) = \mathbf{E}_0 \eta(i+1)$ and $f(i) = b(i,j)$, it follows that

$$
\mathbf{E}_0 x(i+1) = \psi(i,0,\mathbf{E}_0 \eta(\cdot+1)) + \sum_{j=0}^{i} \psi(i,j,b(\cdot,j))\mathbf{E}_0 u(j) + R(i,0)\varphi_0(0).
$$

Putting $i = N-1$, from this and (2.53) we obtain

$$
\begin{aligned}
\mathbf{E}_0 x(N) &= \psi(N-1,0,\mathbf{E}_0 \eta(\cdot+1)) + R(N-1,0)\varphi_0(0) \\
&\quad + \sum_{j=0}^{N-1} \psi(N-1,j,b(\cdot,j))\mathbf{E}_0 u(j) \\
&= \psi(N-1,0,\mathbf{E}_0 \eta(\cdot+1)) + R(N-1,0)\varphi_0(0) \\
&\quad - \sum_{j=0}^{N-1} \psi(N-1,j,b(\cdot,j))G^{-1}(j)\psi'(N-1,j,b(\cdot,j))F\mathbf{E}_0 x_0(N)
\end{aligned}
$$

or

$$
\begin{aligned}
\mathbf{E}_0 x(N) &= \left[I + \sum_{j=0}^{N-1} \psi(N-1,j,b(\cdot,j))G^{-1}(j)\psi'(N-1,j,b(\cdot,j))F\right]^{-1} \\
&\quad \times [\psi(N-1,0,\mathbf{E}_0 \eta(\cdot+1)) + R(N-1,0)\varphi_0(0)].
\end{aligned}
$$

From this and (2.53), (2.57) it follows that $u_0(0)$ has the representation (2.56).

To get $u_0(j)$ for $j > 0$ from (2.47) for $i \geq j \geq 0$ via (2.61), we have

$$\begin{aligned} x_0(i+1) - x_0(j+1) = \eta(i+1) - \eta(j+1) \\ + \sum_{k=0}^{j} a_j(i,k)x_0(k) + \sum_{k=j+1}^{i} a(i,k)x_0(k) \\ + \sum_{k=0}^{j} b_j(i,k)u_0(k) + \sum_{k=j+1}^{i} b(i,k)u_0(k). \end{aligned}$$

So, using (2.59) and putting

$$\zeta(i,j+1) = \begin{cases} x_0(j+1) + \beta(i,j+1) + \sum_{k=0}^{j} a_j(i,k)x_0(k) \\ + \sum_{k=0}^{j} b_j(i,k)u_0(k) + \sum_{k=j+1}^{i} b(i,k)\mathbf{E}_{j+1}u_0(k), & j < i, \\ x_0(i+1), & j = i, \end{cases} \tag{2.65}$$

we have

$$\mathbf{E}_{j+1}x_0(i+1) = \zeta(i,j+1) + \sum_{k=j+1}^{i} a(i,k)\mathbf{E}_{j+1}x_0(k), \qquad i \geq j.$$

From this via the resolvent $R(i,k)$ of the kernel $a(i,k)$, we get

$$\mathbf{E}_{j+1}x_0(i+1) = \zeta(i,j+1) + \sum_{k=j+1}^{i} R(i,k)\zeta(k-1,j+1). \tag{2.66}$$

Substituting (2.65) into (2.66), we obtain

$$\begin{aligned} \mathbf{E}_{j+1}x_0(i+1) = \left[I + \sum_{m=j+1}^{i} R(i,m) \right] x_0(j+1) \\ + \beta(i,j+1) + \sum_{m=j+1}^{i} R(i,m)\beta(m-1,j+1) \\ + \sum_{k=0}^{j} \left[a_j(i,k) + \sum_{m=j+1}^{i} R(i,m)a_j(m-1,k) \right] x_0(k) \end{aligned}$$

$$+\sum_{k=0}^{j}\left[b_j(i,k)+\sum_{m=j+1}^{i}R(i,m)b_j(m-1,k)\right]u_0(k)$$

$$+\sum_{k=j+1}^{i}\left[b(i,k)+\sum_{m=k+1}^{i}R(i,m)b(m-1,k)\right]\mathbf{E}_{j+1}u_0(k)$$

$$=\psi(i,j,I)x_0(j+1)+\psi(i,j,\beta(\cdot,j+1))$$

$$+\sum_{k=0}^{j}\psi(i,j,a_j(\cdot,k))x_0(k)+\sum_{k=0}^{j}\psi(i,j,b_j(\cdot,k))u_0(k)$$

$$+\sum_{k=j+1}^{i}\psi(i,k,b(\cdot,k))\mathbf{E}_{j+1}u_0(k).$$

Putting now $i = N-1$ and

$$\zeta_0(j+1)=\psi(N-1,j,I)x_0(j+1)+\psi(N-1,j,\beta(\cdot,j+1))$$
$$+\sum_{k=0}^{j}\psi(N-1,j,a_j(\cdot,k))x_0(k),\qquad j\geq 0, \tag{2.67}$$

from this we get

$$\mathbf{E}_{j+1}x_0(N)=\zeta_0(j+1)+\sum_{k=0}^{j}\psi(N-1,j,b_j(\cdot,k))u_0(k)$$
$$+\sum_{k=j+1}^{N-1}\psi(N-1,k,b(\cdot,k))\mathbf{E}_{j+1}u_0(k),\qquad j\geq 0. \tag{2.68}$$

Calculating the conditional mathematical expectation of (2.53), for $j<k$ we have

$$\mathbf{E}_{j+1}u_0(k)=-G^{-1}(k)\psi'(N-1,k,b(\cdot,k))F\mathbf{E}_{j+1}x_0(N). \tag{2.69}$$

Substituting (2.69) into (2.68), we obtain

$$\mathbf{E}_{j+1}x_0(N)=\left[I+\sum_{k=j+1}^{N-1}\psi(N-1,k,b(\cdot,k))G^{-1}(k)\psi'(N-1,k,b(\cdot,k))F\right]^{-1}$$
$$\times\left[\zeta_0(j+1)+\sum_{k=0}^{j}\psi(N-1,j,b_j(\cdot,k))u_0(k)\right]. \tag{2.70}$$

Substituting (2.70) into (2.53) and using (2.57), we get

$$u_0(j+1) = w(j+1) + \sum_{k=0}^{j} p(j+1)\psi(N-1, j, b_j(\cdot, k))u_0(k),$$
$$w(j+1) = p(j+1)\zeta_0(j+1), \quad j \geq 0, \quad w(0) = u_0(0). \tag{2.71}$$

So, the optimal control $u_0(j)$ of the control problem (2.47), (2.48) satisfies the Eq. (2.71). Via (1.17) it admits the representation

$$u_0(j+1) = p(j+1)\zeta_0(j+1) + \sum_{k=1}^{j} Q(j,k)p(k)\zeta_0(k) + Q(j,0)u_0(0), \quad j \geq 0, \tag{2.72}$$

where $Q(j,k)$ is the resolvent of the kernel $p(j+1)\psi(N-1, j, b_j(\cdot, k))$. Substituting $\zeta_0(j)$ given in (2.67) into (2.72), we obtain

$$\begin{aligned}
u_0(j+1) = {} & p(j+1)\psi(N-1, j, \beta(\cdot, j+1)) \\
& + p(j+1)\psi(N-1, j, I)x_0(j+1) \\
& + p(j+1)\sum_{k=0}^{j} \psi(N-1, j, a_j(\cdot, k))x_0(k) + Q(j,0)u_0(0) \\
& + \sum_{k=1}^{j} Q(j,k)p(k)\psi(N-1, k-1, \beta(\cdot, k)) \\
& + \sum_{k=1}^{j} Q(j,k)p(k)\psi(N-1, k-1, I)x_0(k) \\
& + \sum_{k=1}^{j} Q(j,k)p(k)\sum_{l=0}^{k-1} \psi(N-1, k-1, a_{k-1}(\cdot, l))x_0(l).
\end{aligned}$$

Changing the order of summation in the last expression and using (2.58), (2.60), we obtain the optimal control $u_0(j+1)$ in the form (2.56) for $j \geq 0$. The proof is completed.

Remark 2.1 Since (2.48) is a quadratic functional (i.e., a convex functional) and the control (2.56) has the bounded second moment, then the necessary condition of optimality is the sufficient condition too.

Remark 2.2 If the process $\eta(i)$ is a martingale with respect to σ-algebra $\mathfrak{F}_i$, i.e., $\mathbf{E}\left(\eta(i)/\mathfrak{F}_j\right) = \eta(j)$, $i > j$, then $\beta(i, j) = 0$, $i \geq j$, and therefore in (2.56) $\alpha(i) = 0$, $i \in Z$.

Remark 2.3 Note that Theorem 2.2 can be easily generalized on the case of the more general performance functional

$$J(u) = \mathbf{E}\left[x'(N)Fx(N) + \sum_{j=0}^{N-1}(x'(j)F_1(j)x(j) + u'(j)G(j)u(j))\right],$$

where F_1 is a positive semidefinite matrix and the matrices F and $G(j)$ satisfy the previous conditions. In particular, the representation (2.53) in this case takes the form

$$\begin{aligned} u_0(j) = &-G^{-1}(j)\Big[\psi'(N-1, j, b(\cdot, j))F\mathbf{E}_j x_0(N) \\ &+ \sum_{i=j}^{N-2}\psi'(i, j, b(\cdot, j))F_1(i+1)\mathbf{E}_j x_0(i+1)\Big], \quad j = 0, \dots, N-1. \end{aligned}$$

Note that the performance functional of such type is considered below in Example 2.5, in Sect. 6.4 and also in Chap. 4 with $F = 0$, $N = \infty$.

2.2.2 Examples

Here, some examples of linear-quadratic optimal control problems of the type of (2.47) and (2.48) are considered.

Example 2.1 Consider the optimal control problem for the scalar linear stochastic difference equation

$$x(i+1) = \eta(i+1) + \sum_{j=0}^{i} a(i, j)x(j) + \sum_{j=0}^{i} b(i, j)u(j), \quad i \geq 0, \quad x(0) = \varphi_0(0), \tag{2.73}$$

with the quadratic performance functional

$$J(u) = \mathbf{E}\left[Fx^2(N) + \sum_{i=0}^{N-1} G(i)u^2(i)\right], \tag{2.74}$$

where $F \geq 0$, $G(i) > 0$, $i = 0, \dots, N-1$. The control problem (2.73), (2.74) is a particular case of the optimal control problem (2.47), (2.48). So, the optimal control of this problem is represented in the form (2.56)–(2.61). In this case, the function $\psi(N-1, j, b(\cdot, j))$ from (2.57) can be represented in some more simple form. Really, rewrite (2.50) for $q_0(i)$ in the matrix form

$$Q_0 = AQ_0 + BV. \tag{2.75}$$

Here, the matrices A and B have the dimension $(N+1) \times (N+1)$, the vectors Q_0 and V have the dimension $N+1$ and, respectively, equal

$$A = \begin{pmatrix} 0 & 0 & \cdots & 0 & 0 \\ a(0,0) & 0 & \cdots & 0 & 0 \\ a(1,0) & a(1,1) & \cdots & 0 & 0 \\ \cdots & \cdots & \cdots & \cdots & \cdots \\ a(N-1,0) & a(N-1,1) & \cdots & a(N-1,N-1) & 0 \end{pmatrix},$$

$$B = \begin{pmatrix} 0 & 0 & \cdots & 0 & 0 \\ b(0,0) & 0 & \cdots & 0 & 0 \\ b(1,0) & b(1,1) & \cdots & 0 & 0 \\ \cdots & \cdots & \cdots & \cdots & \cdots \\ b(N-1,0) & b(N-1,1) & \cdots & b(N-1,N-1) & 0 \end{pmatrix},$$

$$Q_0 = \begin{pmatrix} q_0(0) \\ q_0(1) \\ q_0(2) \\ \cdots \\ q_0(N) \end{pmatrix}, \qquad V = \begin{pmatrix} v(0) \\ v(1) \\ v(2) \\ \cdots \\ v(N) \end{pmatrix}.$$

Since $\det(I - A) = 1$, then there exists the inverse matrix $D = (I - A)^{-1}$, and the solution of the equation (2.75) can be represented in the form

$$Q_0 = DBV. \tag{2.76}$$

Comparing (2.76) with (2.54), we obtain that $\psi(N-1, j, b(\cdot, j)) = r(N, j)$, where $r(N, j)$, $j = 0, 1, \ldots, N$, are elements of the last line of the matrix DB. Then the expression (2.57) can be written in the form

$$p(j) = -\frac{Fr(N,j)}{G(j)} \left[1 + F \sum_{k=j}^{N-1} \frac{r^2(N,k)}{G(k)} \right]^{-1}. \tag{2.77}$$

Remark 2.4 To calculate $r(N, j)$, it is enough to know the last line of the matrix D only.

Example 2.2 Consider the optimal control problem for the scalar linear stochastic difference equation

$$x(i+1) = \eta(i+1) + a \sum_{j=0}^{i} x(j) + b \sum_{j=0}^{i} u(j), \quad i \geq 0, \quad x(0) = \varphi_0(0), \tag{2.78}$$

with the quadratic performance functional

$$J(u) = \mathbf{E}\left[Fx^2(N) + \lambda \sum_{i=0}^{N-1} u^2(i) \right], \tag{2.79}$$

where

$$F > 0, \qquad \lambda > 0, \qquad \eta(i+1) = \varphi_0(0) + \sum_{j=0}^{i} \sigma(j)\xi(j+1), \qquad i \in Z, \tag{2.80}$$

$\sigma(j)$ are arbitrary constants, $\xi(j)$ are $\mathfrak{F}_j$-adapted mutually independent random variables such that $\mathbf{E}\xi(j) = 0$, $\mathbf{E}\xi^2(j) = 1$, $j \in Z$.

In this case, the matrices A, B, and D from Example 2.1 are respectively

$$A = \begin{pmatrix} 0 & 0 & 0 & \cdots & 0 & 0 \\ a & 0 & 0 & \cdots & 0 & 0 \\ a & a & 0 & \cdots & 0 & 0 \\ \cdots & \cdots & \cdots & \cdots & \cdots & \cdots \\ a & a & a & \cdots & a & 0 \end{pmatrix},$$

$$B = \begin{pmatrix} 0 & 0 & 0 & \cdots & 0 & 0 \\ b & 0 & 0 & \cdots & 0 & 0 \\ b & b & 0 & \cdots & 0 & 0 \\ \cdots & \cdots & \cdots & \cdots & \cdots & \cdots \\ b & b & b & \cdots & b & 0 \end{pmatrix},$$

and

$$D = \begin{pmatrix} 1 & 0 & 0 & \cdots & 0 & 0 \\ a & 1 & 0 & \cdots & 0 & 0 \\ a(a+1) & a & 1 & \cdots & 0 & 0 \\ \cdots & \cdots & \cdots & \cdots & \cdots & \cdots \\ a(a+1)^{N-1} & a(a+1)^{N-2} & a(a+1)^{N-3} & \cdots & a & 1 \end{pmatrix}.$$

To calculate the last line of the matrix DB note that

$$\begin{aligned} r(N,j) &= b\left[1 + a + a(a+1) + a(a+1)^2 + \cdots + a(a+1)^{N-2-j}\right] \\ &= b\left[1 + a\left(1 + (a+1) + (a+1)^2 + \cdots + (a+1)^{N-2-j}\right)\right] \\ &= b\left[1 + a\frac{(a+1)^{N-1-j} - 1}{a}\right] \\ &= b(a+1)^{N-1-j}. \end{aligned}$$

So, the last line of the matrix DB is

$$\left(b(a+1)^{N-1},\ b(a+1)^{N-2},\dots,b(a+1),\ b,\ 0\right),$$

i.e.,

$$\begin{aligned} r(N,j) &= b(a+1)^{N-1-j} = b\psi(N-1,j,1),\\ &\quad j=0,1,\dots,N-1, \qquad r(N,N)=0. \end{aligned} \tag{2.81}$$

Let us calculate $p(j)$. Using (2.77), (2.81), we have

$$\begin{aligned} p(j) &= -\frac{Fb}{\lambda}(a+1)^{N-1-j}\left[1+\frac{Fb^2}{\lambda}\sum_{k=j}^{N-1}(a+1)^{2(N-1-k)}\right]^{-1}\\ &= -b(a+1)^{N-1-j}\left[\frac{\lambda}{F}+\frac{b^2}{a(a+2)}\left((a+1)^{2(N-j)}-1\right)\right]^{-1}, \end{aligned} \tag{2.82}$$

$j=0,1,\dots,N-1.$

Via (2.80) $\eta(i)$ is a martingale. So, $\mathbf{E}_0\eta(i)=\varphi_0(0)$, $i\geq 0$, and via Remark 2.2 $\beta(i,j)=0$ and $\alpha(i)=0$, $0\leq j\leq i\leq N-1$. From (2.78), (2.61) it follows that $a_j(i,k)=b_j(i,k)=0$. So, $Q(j,k)=0$ and via (2.60) $\gamma(j,k)=0$, $0\leq k\leq j\leq i\leq N$. The resolvent $R(i,j)$ of the kernel $a(i,j)=a$ depends on one argument only, i.e., $R(i,j)=R(i-j)$, and can be obtained via the recurrence relation (1.20):

$$R(i)=a\left(1+\sum_{j=0}^{i-1}R(j)\right).$$

So,

$$\begin{aligned} R(0) &= a,\\ R(1) &= a(1+a),\\ R(2) &= a[1+a+a(1+a)]\\ &= a(1+a)^2, \\ &\dots\dots\dots\dots\dots\dots\\ R(i) &= a\left[1+a+a(1+a)+\dots+a(1+a)^{i-1}\right]\\ &= a(1+a)^i, \end{aligned} \tag{2.83}$$

i.e., $R(i) = a(1+a)^i$, $i \ge 0$. Besides, via (2.56) the optimal control has the form

$$\begin{aligned} u_0(0) &= p(0)[\psi(N-1,0,1) + R(N-1)]\varphi_0(0), \\ u_0(j) &= p(j)\psi(N-1,j-1,1)x_0(j), \\ j &= 1,\dots,N-1. \end{aligned} \tag{2.84}$$

Note that via (2.52), (2.83)

$$\begin{aligned} \psi(N-1,0,1) + R(N-1) &= 1 + \sum_{k=0}^{N-1} R(N-1-k) \\ &= 1 + a\left(1 + (1+a) + \cdots + (1+a)^{N-1}\right) \\ &= 1 + a\frac{(1+a)^N - 1}{a} \\ &= (1+a)^N \end{aligned}$$

and

$$\begin{aligned} \psi(N-1,j-1,1) &= 1 + \sum_{k=j}^{N-1} R(N-1-k) \\ &= 1 + a\left(1 + (1+a) + \cdots + (1+a)^{N-1-j}\right) \\ &= 1 + a\frac{(1+a)^{N-j} - 1}{a} \\ &= (1+a)^{N-j}. \end{aligned}$$

As a result from this and (2.82), (2.84) we obtain

$$\begin{aligned} u_0(j) &= -q(j)x_0(j), \\ q(j) &= \frac{b(a+1)^{2(N-j)-1}}{\lambda F^{-1} + b^2[a(a+2)]^{-1}[(a+1)^{2(N-j)} - 1]}, \\ j &= 0,\dots,N-1. \end{aligned} \tag{2.85}$$

If, for example, $N = 1$, then via (2.85) we have

$$u_0(0) = -\frac{b(a+1)x(0)}{\lambda F^{-1} + b^2}. \tag{2.86}$$

Remark 2.5 Let us show that the control (2.86) can be obtained immediately from the optimal control problem (2.78)–(2.80). Really, from (2.78) for $i = 0$ we have

$$x(1) = \eta(1) + ax(0) + bu(0).$$

Substituting this into (2.79), by $N = 1$ we obtain

$$\begin{aligned}J(u) &= \mathbf{E}[Fx^2(1) + \lambda u^2(0)] \\ &= F\mathbf{E}[(\eta(1) + ax(0) + bu(0))^2 + \lambda F^{-1}u^2(0)] \\ &= F\mathbf{E}[\mathbf{E}_0(\eta(1) + ax(0))^2 + 2b(\mathbf{E}_0\eta(1) + ax(0))u(0) + (\lambda F^{-1} + b^2)u^2(0)].\end{aligned}$$

Note now that via (2.80) $\mathbf{E}_0\eta(1) = \varphi_0(0)$ and a minimum in the square brackets is reached for $u(0) = u_0(0)$ given by (2.86).

Remark 2.6 Note that in the considered optimal control problem (2.78)–(2.80) the obtained optimal control $u_0(i)$ does not depend on the past, i.e., on $x_0(j)$ for $j < i$. Really, it is not a surprise since from (2.78), (2.80) it follows that

$$x(i+1) - x(i) = ax(i) + bu(i) + \sigma(i)\xi(i+1), \qquad i \geq 0. \tag{2.87}$$

So, the initial equation (2.78) is an analogue of the Ito stochastic differential equation without delay

$$dx(t) = (ax(t) + bu(t))dt + \sigma(t)dw(t),$$

where $w(t)$ is the standard Wiener process, $\xi(i+1) = w(t_{i+1}) - w(t_i)$, $x(i) = x(t_i)$, $t_{i+1} - t_i = 1$.

Example 2.3 Let us show that a small modification of the optimal control $u_0(i)$ of the control problem (2.78)–(2.80) leads to an increase of the performance functional.

For this aim, note that via (2.85), (2.87) for $i = N - 1$ we obtain

$$x_0(N) = (1 + a - bq(N-1))x_0(N-1) + \sigma(N-1)\xi(N). \tag{2.88}$$

Using (2.85), (2.88) and the properties of the process $\xi(i)$, represent the performance functional (2.79) in the form

$$\begin{aligned}J(u_0) &= \mathbf{E}\left[Fx_0^2(N) + \lambda\sum_{i=0}^{N-1} q^2(i)x_0^2(i)\right] \\ &= Q\mathbf{E}x_0^2(N-1) + \sigma^2(N-1) + \lambda\sum_{i=0}^{N-2} q^2(i)\mathbf{E}x_0^2(i),\end{aligned}$$

where

$$Q = F(1 + a - bq(N-1))^2 + \lambda q^2(N-1). \tag{2.89}$$

Note that via (2.85)

$$q(N-1) = \frac{b(1+a)}{\lambda F^{-1} + b^2}.$$

Substituting this into (2.89), we obtain

$$\begin{aligned} Q &= F\left(1 + a - \frac{b^2(1+a)}{\lambda F^{-1} + b^2}\right)^2 + \frac{\lambda b^2(1+a)^2}{(\lambda F^{-1} + b^2)^2} \\ &= (1+a)^2\left[F\left(\frac{\lambda F^{-1}}{\lambda F^{-1} + b^2}\right)^2 + \frac{\lambda b^2}{(\lambda F^{-1} + b^2)^2}\right] \\ &= \frac{\lambda(1+a)^2}{\lambda F^{-1} + b^2}. \end{aligned}$$

Put now, for example,

$$q(N-1) = \frac{1+a}{b}, \tag{2.90}$$

and let all other $q(i)$ let be defined by (2.85). In this case for arbitrary $F > 0$, we obtain

$$Q = \frac{\lambda(1+a)^2}{b^2} > \frac{\lambda(1+a)^2}{\lambda F^{-1} + b^2},$$

i.e., the performance functional (2.79) is increasing.

For numerical simulation, put

$$\begin{aligned} &N = 10, \qquad a = 0.5, \qquad b = 0.1, \qquad F = \lambda = 1, \\ &x(0) = 3, \qquad \sigma(j) = 0, \qquad j = 0, \ldots, N-1. \end{aligned} \tag{2.91}$$

In Fig 2.1 one can see the trajectories of $q(j)$ defined in (2.85), of the optimal solution $x_0(j)$ and of the optimal control $u_0(j)$ given in (2.85), (2.85), respectively, by the values of the parameters (2.91). The minimal value of the performance functional is $J(u_0) = 1{,}085$.

Changing $q(j)$ in the one point $i = N - 1$ only as it is given in (2.90), we obtain another picture (see Fig. 2.2). Via (2.88), (2.91) here $x(N) = 0$ but $\bar{u}(N-1) = -63.32$ and $J(\bar{u}) = 5{,}054 > J(u_0)$.

Consider also the case $q(0) = (1+a)b^{-1}$. By that via (2.87), (2.91), we obtain (see Fig. 2.3) $\bar{u}(j) = x(j) = 0$, $j > 0$, and $J(\bar{u}) = 2{,}025 > J(u_0)$.

Example 2.4 Consider the optimal control problem for the scalar linear stochastic difference equation

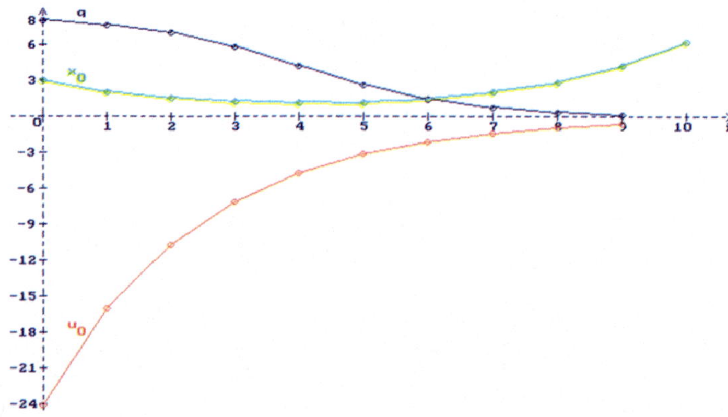

Fig. 2.1 $J(u_0) = 1{,}085$

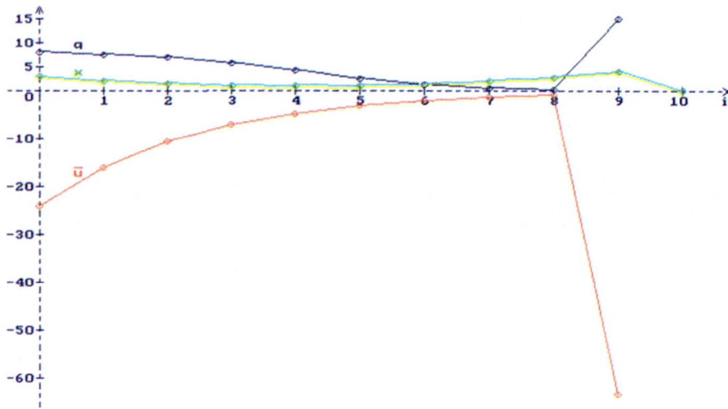

Fig. 2.2 $J(\bar{u}) = 5{,}054$

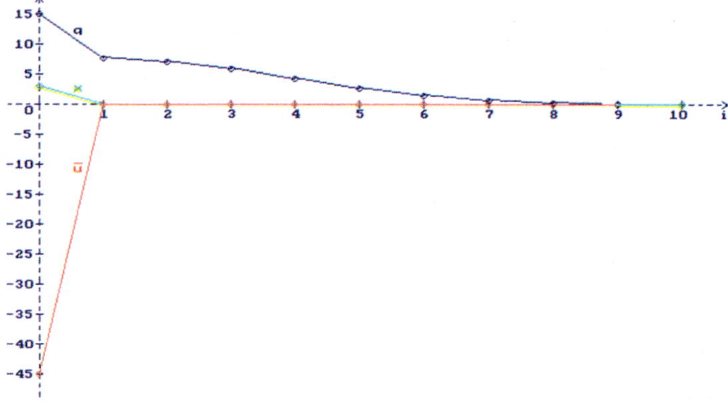

Fig. 2.3 $J(\bar{u}) = 2{,}025$

$$\begin{aligned} x(i+1) &= \eta(i+1) + a_0 x(i) + a_1 x(i-1) + bu(i), \\ i \geq 0, \quad & x(0) = \varphi_0(0) = \eta(0), \quad x(-1) = \varphi_0(-1), \end{aligned} \tag{2.92}$$

where $\eta(i)$ is a martingale, and the quadratic performance functional

$$J(u) = \mathbf{E}\left[x^2(N) + \lambda \sum_{i=0}^{N-1} u^2(i) \right]. \tag{2.93}$$

The Eq. (2.92) can be represented in the form (2.47) as follows:

$$\begin{aligned} x(1) &= \hat{\eta}(1) + a_0 x(0) + bu(0), \\ x(i+1) &= \hat{\eta}(i+1) + a_0 x(i) + a_1 x(i-1) + bu(i), \\ i &\geq 1, \quad x(0) = \varphi_0(0), \end{aligned}$$

where

$$\hat{\eta}(i) = \begin{cases} \eta(i), & i > 1, \\ \eta(1) + a_1 \varphi_0(-1), & i = 1. \end{cases} \tag{2.94}$$

Note that via (1.21) and (2.92) the resolvent $R(i)$ is

$$\begin{aligned} R(0) &= a_0, \\ R(1) &= a_0 R(0) + a_1 \\ &= a_1 + a_0^2, \\ R(2) &= a_0 R(1) + a_1 R(0) \\ &= 2a_0 a_1 + a_0^3, \\ R(3) &= a_0 R(2) + a_1 R(1) \\ &= a_1^2 + 3a_0^2 a_1 + a_0^4, \\ &\dots\dots\dots\dots\dots\dots \\ R(i) &= a_0 R(i-1) + a_1 R(i-2), \quad i \geq 2. \end{aligned} \tag{2.95}$$

Note also that via (2.52), (2.57), (2.92)

$$\begin{aligned} \psi(N-1, j, b(\cdot, j)) &= \left(1 + \sum_{k=j+1}^{N-1} R(N-1-k) \right) b \\ &= \begin{cases} \psi(N-1, j, 1)b, & j = 0, 1, \dots, N-2, \\ b, & j = N-1, \end{cases} \end{aligned}$$

and

$$p(j) = \begin{cases} -\frac{\psi(N-1,j,1)b}{\lambda+b^2\left(1+\sum\limits_{k=j}^{N-2}\psi^2(N-1,k,1)\right)}, & j = 0, 1, \dots, N-2, \\ -\frac{b}{\lambda+b^2}, & j = N-1, \end{cases} \tag{2.96}$$

Let be $N = 1$. Using that $\eta(i)$ is a martingale and (2.94), we have

$$\begin{aligned} \psi(0, 0, \mathbf{E}_0\hat{\eta}(\cdot+1)) &= \mathbf{E}_0\hat{\eta}(1) \\ &= \mathbf{E}_0(\eta(1) + a_1\varphi_0(-1)) \\ &= \varphi_0(0) + a_1\varphi_0(-1). \end{aligned}$$

As a result from (2.56), (2.95), (2.96) we obtain

$$u_0(0) = -\frac{b}{\lambda+b^2}\left[(1+a_0)\varphi_0(0) + a_1\varphi_0(-1)\right].$$

Let be $N \geq 2$. Then

$$\begin{aligned} \psi(N-1, 0, \mathbf{E}_0\hat{\eta}(\cdot+1)) &= \mathbf{E}_0\hat{\eta}(N) + \sum_{k=1}^{N-1} R(N-1-k)\mathbf{E}_0\hat{\eta}(k) \\ &= \mathbf{E}_0\eta(N) + \sum_{k=2}^{N-1} R(N-1-k)\mathbf{E}_0\eta(k) \\ &\quad + R(N-2)\mathbf{E}_0\hat{\eta}(1) \\ &= \left(1 + \sum_{k=2}^{N-1} R(N-1-k)\right)\varphi_0(0) \\ &\quad + R(N-2)(\varphi_0(0) + a_1\varphi_0(-1)) \\ &= \left(1 + \sum_{k=1}^{N-1} R(N-1-k)\right)\varphi_0(0) \\ &\quad + R(N-2)a_1\varphi_0(-1) \\ &= \psi(N-1, 0, 1)\varphi_0(0) + R(N-2)a_1\varphi_0(-1). \end{aligned} \tag{2.97}$$

So, via (2.56), (2.97) for $N \geq 2$ we get

$$u_0(0) = p(0)\left[\psi(N-1, 0, 1)\varphi_0(0) + R(N-2)a_1\varphi_0(-1)\right].$$

For example, for $N = 2$ we have

$$u_0(0) = -\frac{a_0 b}{\lambda + b^2(1 + a_0^2)} \left[\left(1 + a_0 + a_1 + a_0^2\right) \varphi_0(0) + a_0 a_1 \varphi_0(-1) \right].$$

To calculate $u_0(1)$ for $N = 2$ note that via (2.96) and (2.57)–(2.61)

$$\begin{aligned} p(1) &= -\frac{b}{\lambda + b^2}, \qquad \psi(1, 0, 1) = 1 + a_0, \\ \alpha(1) &= p(1)\psi(1, 0, \beta(\cdot, 1)) = p(1)\beta(1, 1) = -p(1)\varphi_0(-1), \\ \gamma(0, 0) &= p(1)\psi(1, 0, a_0(\cdot, 0)) = p(1)a_0(1, 0) = p(1)(a_1 - a_0), \\ Q(0, 0) &= p(1)\psi(1, 0, b_0(\cdot, 0)) = p(1)b_0(1, 0) = -p(1)b. \end{aligned}$$

So, via (2.56) we obtain

$$u_0(1) = -\frac{b}{\lambda + b^2}[(1 + a_0)x_0(1) + (a_1 - a_0)\varphi_0(0) - \varphi_0(-1) - bu_0(0)].$$

Similarly, one can calculate the optimal control $u_0(0), u_0(1), \ldots, u_0(N - 1)$ of the problem (2.92), (2.93) for arbitrary N.

Example 2.5 Let us show that similarly to the control problem (2.47), (2.48) the optimal control can be obtained for linear-quadratic problems with linear equations that are different from the equation (2.47) and the quadratic performance functional that was considered in Remark 2.3. Consider, for example, the scalar controlled process with a noise in control

$$\begin{aligned} x(i + 1) = \eta(i + 1) \quad &\sum_{j=0}^{i} [\beta(i, j) + \gamma(i, j)u(j)]\,\xi(j + 1), \\ &x(0) = \varphi_0(0), \end{aligned} \tag{2.98}$$

and the quadratic performance functional

$$J(u) = \mathbf{E}\left[x^2(N) + \sum_{i=0}^{N-1} (\lambda_0 x^2(i) + \lambda_1 u^2(i)) \right]. \tag{2.99}$$

Here, $\lambda_0 \geq 0$, $\lambda_1 > 0$, $\eta(i) \in H$, $\xi(j)$, $j \in Z$, are $\mathfrak{F}_j$-adapted Gaussian random variables that are mutually independent and are independent on $\eta(i)$, $\mathbf{E}\xi(j) = 0$, $\mathbf{E}\xi^2(j) = 1$.

Via (2.98), (2.99) in this case

$$q_0(i+1) = \sum_{j=0}^{i} \gamma(i,j)v(j)\xi(j+1),$$

$$J_0(u_0) = 2\mathbf{E}\left[x_0(N)q_0(N) + \sum_{i=0}^{N-1}(\lambda_0 x_0(i)q_0(i) + \lambda_1 u_0(i)v(i))\right]. \quad (2.100)$$

Note that $\mathbf{E}\eta(i)\xi(j) = 0$, $i, j \in Z$, and $\mathbf{E}\xi(i)\xi(j) = 0$, $i \neq j$. So, via (2.98), (2.100) we have

$$\begin{aligned}\mathbf{E}x_0(i)q_0(i) &= \mathbf{E}\left[\eta(i) + \sum_{j=0}^{i-1}[\beta(i-1,j) + \gamma(i-1,j)u_0(j)]\xi(j+1)\right] \\ &\quad \times \sum_{l=0}^{i-1}\gamma(i-1,l)v(l)\xi(l+1) \\ &= \sum_{j=0}^{i-1}\mathbf{E}\,[\beta(i-1,j) + \gamma(i-1,j)u_0(j)]\,\gamma(i-1,j)v(j)\end{aligned}$$

and

$$\begin{aligned}\sum_{i=0}^{N-1}\mathbf{E}x_0(i)q_0(i) &= \sum_{i=0}^{N-1}\sum_{j=0}^{i-1}\mathbf{E}\,[\beta(i-1,j) + \gamma(i-1,j)u_0(j)]\,\gamma(i-1,j)v(j) \\ &= \sum_{j=0}^{N-2}\sum_{i=j+1}^{N-1}\mathbf{E}\,[\beta(i-1,j) + \gamma(i-1,j)u_0(j)]\,\gamma(i-1,j)v(j)\end{aligned}$$

From this and (2.100), it follows that

$$\begin{aligned}J_0(u_0) = 2\mathbf{E}\sum_{j=0}^{N-1}&\Bigg[[\beta(N-1,j) + \gamma(N-1,j)u_0(j)]\gamma(N-1,j) + \lambda_1 u_0(j) \\ &+ \lambda_0 \sum_{i=j+1}^{N-1}[\beta(i-1,j) + \gamma(i-1,j)u_0(j)]\,\gamma(i-1,j)\Bigg]v(j).\end{aligned}$$

Thus, the necessary condition for optimality $J_0(u_0) \geq 0$ of the control u_0 holds if and only if the expression in the square brackets equals zero, i.e.,

$$\begin{aligned}&[\beta(N-1,j) + \gamma(N-1,j)u_0(j)]\,\gamma(N-1,j) + \lambda_1 u_0(j) \\ &\quad + \lambda_0 \sum_{i=j+1}^{N-1}[\beta(i-1,j) + \gamma(i-1,j)u_0(j)]\,\gamma(i-1,j) = 0.\end{aligned}$$

From this, it follows that the optimal control u_0 of the control problem (2.98), (2.99) has the form

$$u_0(j) = -\frac{\beta(N-1, j)\gamma(N-1, j) + \lambda_0 \sum_{i=j+1}^{N-1} \beta(i-1, j)\gamma(i-1, j)}{\lambda_1 + \gamma^2(N-1, j) + \lambda_0 \sum_{i=j+1}^{N-1} \gamma^2(i-1, j)},$$

$$j = 0, 1, \ldots, N-1.$$

Chapter 3
Successive Approximations to the Optimal Control

In this section, successive approximations to the optimal control of a quasilinear stochastic difference equation with quadratic performance functional are constructed. The algorithm of construction of successive approximations is based on the technique of construction of some auxiliary control problem.

3.1 Statement of the Problem

Consider the optimal control problem for stochastic quasilinear difference equation

$$\begin{aligned} x(i+1) &= \eta(i+1) + \sum_{j=0}^{i} a(i,j)x(j) + \sum_{j=0}^{i} b(i,j)u(j) \\ &\quad + \varepsilon \left[\Phi(i+1, x_{i+1}) + \sum_{j=0}^{i} \sigma(i,j,x_j)\xi(j+1) \right], \qquad (3.1) \\ i &= 0, \ldots, N-1, \quad x(j) = \varphi_0(j), \quad j \in Z_0 = [-h, \ldots, 0], \end{aligned}$$

with quadratic performance functional

$$J(u) = \mathbf{E}\left[x'(N)Fx(N) + \sum_{j=0}^{N-1} u'(j)G(j)u(j) \right]. \qquad (3.2)$$

Here $\varepsilon \geq 0$, $\eta \in H$, $\mathfrak{F}_i$-adapted random variables $\xi(i) \in \mathbf{R}^l$ are mutually independent on each other and on the $\mathfrak{F}_i$-adapted process $\eta(i)$, $\mathbf{E}\xi(i) = 0$, $\mathbf{E}\xi(i)\xi'(i) = I$, where I is the identical matrix, x_i is a trajectory of the process x until the moment of time i, $a(i,j)$, $b(i,j)$ and $\sigma(i,j,\varphi)$ are nonrandom matrices of the dimension $n \times n$, $n \times m$ and $n \times l$, respectively, F is a positive semidefinite $n \times n$-matrix, $G(j)$, $j = 0, 1, \ldots, N-1$, is a positive definite $l \times l$-matrix.

L. Shaikhet, *Optimal Control of Stochastic Difference Volterra Equations*,
Studies in Systems, Decision and Control 17, DOI 10.1007/978-3-319-13239-6_3

The functional $\Phi(i, \varphi) \in \mathbf{R}^n$ depends on the values of the function $\varphi(j)$ for $j = -h, -h+1, \ldots, i$, $\varphi \in \tilde{H}$, and satisfies the conditions: $\Phi(i, 0) = 0$,

$$|\Phi(i, \varphi)| \leq \sum_{j=-h}^{i} (1 + |\varphi(j)|)\, K_1(j), \tag{3.3}$$

for arbitrary functions $\varphi_1, \varphi_2 \in \tilde{H}$

$$|\Phi(i, \varphi_1) - \Phi(i, \varphi_2)| \leq \sum_{j=-h}^{i} |\varphi_1(j) - \varphi_2(j)|\, K_1(j), \tag{3.4}$$

the $n \times l$-matrix $\sigma(i, j, \varphi)$ depends on values of the function $\varphi(l)$ for $l = -h, -h+1, \ldots, j$, $\varphi \in \tilde{H}$, and satisfies the conditions

$$|\sigma(i, j, \varphi)|^2 \leq \sum_{l=-h}^{j} (1 + |\varphi(l)|^2) K_1(l), \tag{3.5}$$

for arbitrary functions $\varphi_1, \varphi_2 \in \tilde{H}$

$$|\sigma(i, j, \varphi_1) - \sigma(i, j, \varphi_2)|^2 \leq \sum_{l=-h}^{j} |\varphi_1(l) - \varphi_2(l)|^2\, K_1(l). \tag{3.6}$$

It is supposed that $K_1(l)$ in the conditions (3.3)–(3.6) satisfies the condition (2.13).

Definition 3.1 Put $V = \inf_{u \in \mathbf{U}} J(u)$. An admissible control u_k, $k = 0, 1, \ldots$, such that

$$0 \leq J(u_k) - V \leq C\varepsilon^{k+1} \tag{3.7}$$

for some positive C and ε, is called kth approximation to the optimal control of the control problem (3.1), (3.2).

Below a sequence of kth successive approximations, $k = 0, 1, \ldots$, to the optimal control of the control problem (3.1), (3.2) is constructed.

3.2 Algorithm of Successive Approximations Construction

Let us construct a sequence $(x_k(i), u_k(i))$, $i \in Z$, $k = 0, 1, \ldots$, by the following way. Let $(x_0(i), u_0(i))$ be the optimal trajectory and the optimal control of the control problem with the equation

$$x(i+1) = \eta(i+1) + \sum_{j=0}^{i} a(i,j)x(j) + \sum_{j=0}^{i} b(i,j)u(j), \tag{3.8}$$
$$x(0) = \varphi_0(0),$$

and the performance functional (3.2), and let $(x_k(i), u_k(i))$, $k = 1, 2, \ldots$, be the optimal trajectory and the optimal control of the control problem with the equation

$$\begin{aligned} x(i+1) &= \eta(i+1) + \sum_{j=0}^{i} a(i,j)x(j) + \sum_{j=0}^{i} b(i,j)u(j) \\ &\quad + \varepsilon\left[\Phi(i+1, x_{k-1,i+1}) + \sum_{j=0}^{i} \sigma(i,j,x_{k-1,j})\xi(j+1)\right], \end{aligned} \tag{3.9}$$
$$x(0) = \varphi_0(0), \qquad x_{k-1}(j) = \varphi_0(j), \quad j \in Z_0,$$

and the performance functional (3.2). Here $x_{k-1}(j)$ for $j = 0, 1, \ldots, i$ is the optimal trajectory of the previous optimal control problem.

The optimal control problems (3.8), (3.2), and (3.9), (3.2) are linear-quadratic problems. From Theorem 2.2 and (2.56) it follows that the optimal control $u_k(j)$, $k = 0, 1, \ldots$, in this case has the form

$$\begin{aligned} u_k(0) &= p(0)\left[\psi(N-1, 0, \mathbf{E}_0\eta_k(\cdot+1)) + R(N-1,0)\varphi_0(0)\right], \\ u_k(j+1) &= \alpha_k(j+1) + p(j+1)\psi(N-1, j, I)x_k(j+1) \\ &\quad + \sum_{l=0}^{j} \gamma(j,l)x_k(l) + Q(j,0)u_k(0), \quad j = 0, 1, \ldots, N-2. \end{aligned} \tag{3.10}$$

Here

$$\begin{aligned} \alpha_k(j+1) &= p(j+1)\psi(N-1, j, \beta_k(\cdot, j+1)) \\ &\quad + \sum_{l=1}^{j} Q(j,l)p(l)\psi(N-1, l-1, \beta_k(\cdot, l)), \quad k = 0, 1, \ldots, \end{aligned} \tag{3.11}$$

$$\eta_0(i) = \eta(i), \qquad \beta_0(i,j) = \mathbf{E}_j\eta(i+1) - \eta(j),$$
$$\mathbf{E}_0\eta_k(i) = \mathbf{E}_0\left[\eta(i) + \varepsilon\Phi(i, x_{k-1,i})\right], \qquad k = 1, 2, \ldots,$$
$$\begin{aligned} \beta_k(i,j) &= \beta_0(i,j) + \varepsilon\Bigg[\mathbf{E}_j\Phi(i+1, x_{k-1,i+1}) - \Phi(j, x_{k-1,j}) + \sum_{l=0}^{j-1}(\sigma(i,l,x_{k-1,l}) \\ &\quad - \sigma(j-1, l, x_{k-1,l}))\xi(l+1)\Bigg], \end{aligned} \tag{3.12}$$

the functional $\psi(i, j, f(\cdot))$ is defined in (2.52), where $R(i, l)$ is the resolvent of the kernel $a(i, l)$, $p(j)$, and $\gamma(j, k)$ are defined as in (2.57) and (2.60), (2.61) respectively, i.e.,

$$
\begin{aligned}
p(j) = & - G^{-1}(j)\psi'(N-1, j, b(\cdot, j))F \\
& \times \left[I + \sum_{k=j}^{N-1} \psi(N-1, k, b(\cdot, k))G^{-1}(k)\psi'(N-1, k, b(\cdot, k))F \right]^{-1}, \\
& j = 0, 1, \ldots, N-1,
\end{aligned}
\tag{3.13}
$$

and

$$
\gamma(j, k) = \begin{cases} p(j+1)\psi(N-1, j, a_j(\cdot, k)) + Q(j, k)p(k)\psi(N-1, k-1, I) \\ + \sum_{l=k+1}^{j} Q(j, l)p(l)\psi(N-1, l-1, a_{l-1}(\cdot, k)), \quad j \geq k \geq 1, \\ p(j+1)\psi(N-1, j, a_j(\cdot, 0)) \\ + \sum_{l=1}^{j} Q(j, l)p(l)\psi(N-1, l-1, a_{l-1}(\cdot, 0)), \quad j \geq k = 0, \end{cases}
\tag{3.14}
$$

where $Q(j, k)$ is the resolvent of the kernel $p(j+1)\psi(N-1, j, b_j(\cdot, k))$,

$$
\begin{aligned}
a_j(i, k) &= a(i, k) - a(j, k), \\
b_j(i, k) &= b(i, k) - b(j, k), \\
0 \leq k &\leq j \leq i \leq N-1.
\end{aligned}
$$

Theorem 3.1 *The sequence $(x_k(i), u_k(i))$, $(k = 0, 1, \ldots)$, of the optimal solutions of the control problems (3.8), (3.2) and (3.9), (3.2) has a limit $(\tilde{x}(i), \tilde{u}(i))$. The processes $\tilde{x}(i)$ and $\tilde{u}(i)$ are defined by the equations*

$$
\begin{aligned}
\tilde{x}(i+1) = & \ \eta(i+1) + \sum_{j=0}^{i} a(i, j)\tilde{x}(j) + \sum_{j=0}^{i} b(i, j)\tilde{u}(j) \\
& + \varepsilon \left[\Phi(i+1, \tilde{x}_{i+1}) + \sum_{j=0}^{i} \sigma(i, j, \tilde{x}_j)\xi(j+1) \right], \\
& i = 0, 1, \ldots, N-1, \quad \tilde{x}(j) = \varphi_0(j), \quad j \in Z_0,
\end{aligned}
\tag{3.15}
$$

$$
\begin{aligned}
\tilde{u}(0) &= p(0)\left[\psi(N-1, 0, \mathbf{E}_0\tilde{\eta}(\cdot+1)) + R(N-1, 0)\varphi_0(0)\right], \\
\tilde{u}(j+1) &= \alpha(j+1) + p(j+1)\psi(N-1, j, I)\tilde{x}(j+1) \\
& \quad + \sum_{l=0}^{j} \gamma(j, l)\tilde{x}(l) + Q(j, 0)\tilde{u}(0),
\end{aligned}
\tag{3.16}
$$

$$\alpha(j+1) = p(j+1)\psi(N-1, j, \beta(\cdot, j+1)) \\ + \sum_{l=0}^{j} Q(j,l)p(l)\psi(N-1, l-1, \beta(\cdot, l)), \tag{3.17}$$

where $j = 0, \ldots, N-2$,

$$\mathbf{E}_0\tilde{\eta}(i) = \mathbf{E}_0\left[\eta(i) + \varepsilon\Phi(i, \tilde{x}_i)\right],$$

$$\beta(i,j) = \beta_0(i,j) + \varepsilon\Bigg[\mathbf{E}_j\Phi(i+1, \tilde{x}_{i+1}) - \Phi(j, \tilde{x}_j) \\ + \sum_{l=0}^{j-1}\Big(b(i,l,\tilde{x}_l) - b(j-1,l,\tilde{x}_l)\Big)\xi(l+1)\Bigg], \tag{3.18}$$

and $p(j)$, $\gamma(j,k)$ *are defined by* (3.13), (3.14).

Proof Let us prove that the sequence $(x_k(i), u_k(i))$ is uniformly bounded with respect to $k = 0, 1, \ldots$, i.e., $\|x_k\|_N \le C$, $\|u_k\|_N \le C$. From (3.9) for $x = x_k$, $u = u_k$, using (3.3), we get

$$|x_k(i+1)| \le |\eta(i+1)| + \sum_{j=0}^{i}|a(i,j)||x_k(j)| + \sum_{j=0}^{i}|b(i,j)||u_k(j)| \\ + \varepsilon\left[\sum_{j=-h}^{i+1}(1 + |x_{k-1}(j)|)K_1(j) + \left|\sum_{j=0}^{i}\sigma(i,j,x_{k-1,j})\xi(j+1)\right|\right], \\ i = 0, 1, \ldots, N-1.$$

Squaring the both parts of the obtained inequality, calculating the expectation and using (1.30), we get

$$\mathbf{E}|x_k(i+1)|^2 \le 6\Bigg[\mathbf{E}|\eta(i+1)|^2 + \sum_{l=0}^{i}|a(i,l)|\sum_{j=0}^{i}|a(i,j)|\mathbf{E}|x_k(j)|^2 \\ + \sum_{l=0}^{i}|b(i,l)|\sum_{j=0}^{i}|b(i,j)|\mathbf{E}|u_k(j)|^2 \\ + \varepsilon^2\Bigg(\Bigg(\sum_{j=-h}^{i+1}K_1(j)\Bigg)^2 + \sum_{l=-h}^{i+1}K_1(l)\sum_{j=-h}^{i+1}\mathbf{E}|x_{k-1}(j)|^2K_1(j) \\ + \sum_{j=0}^{i}\mathbf{E}|\sigma(i,j,x_{k-1,j})|^2\Bigg)\Bigg]. \tag{3.19}$$

Note that via (3.5)

$$\sum_{j=0}^{i}\mathbf{E}|\sigma(i,j,x_{k-1,j})|^2 \le \sum_{j=0}^{i}\sum_{l=-h}^{j}(1+\mathbf{E}|x_{k-1}(l)|^2)K_1(l)$$
$$\le N\sum_{l=-h}^{N}K_1(l)+N\sum_{l=-h}^{N}\mathbf{E}|x_{k-1}(l)|^2K_1(l) \qquad (3.20)$$

and via (2.1), (2.2)

$$\sum_{l=-h}^{N}\mathbf{E}|x_{k-1}(l)|^2K_1(l) \le \sum_{l=-h}^{0}\mathbf{E}|\varphi_0(l)|^2K_1(l)+\sum_{l=1}^{N}\mathbf{E}|x_{k-1}(l)|^2K_1(l)$$
$$\le (\|\varphi_0\|^2+\|x_{k-1}\|_N^2)\sum_{l=-h}^{N}K_1(l). \qquad (3.21)$$

So, putting

$$a=\max_{0\le j\le i\le N-1}|a(i,j)|,\qquad b=\max_{0\le j\le i\le N-1}|b(i,j)|,\qquad \mu=\sum_{l=-h}^{N}K_1(l),$$

and using (3.20), (3.21), from (3.19) we obtain

$$\mathbf{E}|x_k(i+1)|^2 \le 6\Bigg[\|\eta\|_N^2+a^2N\left(\mathbf{E}|\varphi_0(0)|^2+\sum_{j=1}^{i}\mathbf{E}|x_k(j)|^2\right)$$
$$+b^2N\left(\mathbf{E}|u_k(0)|^2+\sum_{j=1}^{i}\mathbf{E}|u_k(j)|^2\right)$$
$$+\varepsilon^2\mu(\mu+N)(1+\|\varphi_0\|^2+\|x_{k-1}\|_N^2)\Bigg].$$

From this for some $C>0$ it follows that

$$\mathbf{E}|x_k(i+1)|^2 \le C\left(1+\varepsilon^2\|x_{k-1}\|_N^2+\sum_{j=1}^{i}\mathbf{E}|u_k(j)|^2+\sum_{j=1}^{i}\mathbf{E}|x_k(j)|^2\right).$$

Via Lemma 1.2 it means that

$$\mathbf{E}|x_k(i+1)|^2 \le C_1 \left(1 + \varepsilon^2 \|x_{k-1}\|_N^2 + \sum_{j=1}^{i} \mathbf{E}|u_k(j)|^2\right). \tag{3.22}$$

where $C_1 = C(1+C)^{N-1}$.

Similarly, from (3.10) one can get that

$$\mathbf{E}|u_k(j)|^2 \le C\left(1 + \mathbf{E}|\alpha_k(j)|^2 + \|x_k\|_j^2\right), \qquad j = 1, 2, \ldots, N-1. \tag{3.23}$$

To estimate $\mathbf{E}|\alpha_k(j)|^2$ note that via (3.12) and (3.3) we have

$$\begin{aligned}
|\beta_k(i,j)| &\le |\beta_0(i,j)| + \varepsilon\Bigg(\mathbf{E}_j|\Phi(i+1, x_{k-1,i+1})| + |\Phi(j, x_{k-1,j})| \\
&\quad + \left|\sum_{l=0}^{j-1} \sigma(i,l,x_{k-1,l})\xi(l+1)\right| \\
&\quad + \left|\sum_{l=0}^{j-1} \sigma(j-1,l,x_{k-1,l})\xi(l+1)\right|\Bigg) \\
&\le |\beta_0(i,j)| + \varepsilon\Bigg(\sum_{l=0}^{i+1}\left(1 + \mathbf{E}_j|x_{k-1}(l)|\right)K_1(l) \\
&\quad + \sum_{l=0}^{j}\left(1 + |x_{k-1}(l)|\right)K_1(l) \\
&\quad + \left|\sum_{l=0}^{j-1} \sigma(i,l,x_{k-1,l})\xi(l+1)\right| \\
&\quad + \left|\sum_{l=0}^{j-1} \sigma(j-1,l,x_{k-1,l})\xi(l+1)\right|\Bigg).
\end{aligned}$$

From this, using the properties of the process $\xi(j)$ and the condition (3.5), it is easy to get that for some $C > 0$

$$\mathbf{E}|\beta_k(i,j)|^2 \le C\left(1 + \varepsilon^2 \|x_{k-1}\|_N^2\right).$$

Via (3.11) the similar estimation holds for $\mathbf{E}|\alpha_k(j)|^2$ too, i.e.,

$$\mathbf{E}|\alpha_k(j)|^2 \le C\left(1 + \varepsilon^2 \|x_{k-1}\|_N^2\right). \tag{3.24}$$

From (3.22)–(3.24) it follows that for some $C > 0$

$$\mathbf{E}|u_k(j)|^2 \le C\left(1 + \varepsilon^2\|x_{k-1}\|_N^2 + \|x_k\|_j^2\right),$$
$$\mathbf{E}|x_k(i+1)|^2 \le C\left(1 + \varepsilon^2\|x_{k-1}\|_N^2 + \sum_{j=1}^{i}\|x_k\|_j^2\right) \qquad (3.25)$$

and via Lemma 1.2

$$\|x_k\|_{i+1}^2 \le C(1 + \varepsilon^2\|x_{k-1}\|_N^2), \qquad i = 0, 1, \ldots, N-1. \qquad (3.26)$$

From (3.26) for small enough $\varepsilon > 0$ (such that $C\varepsilon^2 < 1$) we have

$$\begin{aligned}\|x_k\|_{i+1}^2 &\le C\left[1 + C\varepsilon^2\left(1 + \varepsilon^2\|x_{k-2}\|_N^2\right)\right]\\ &\le C\left[1 + C\varepsilon^2 + C^2\varepsilon^4\left(1 + \varepsilon^2\|x_{k-3}\|_N^2\right)\right]\\ &\le \ldots\\ &\le C\left[1 + C\varepsilon^2 + \cdots + (C\varepsilon^2)^{k-1} + (C\varepsilon^2)^{k-1}\varepsilon^2\|x_0\|_N^2\right]\\ &\le \frac{C}{1 - C\varepsilon^2} + \|x_0\|_N^2\\ &< \infty.\end{aligned}$$

So, the sequence $\|x_k\|_i$ is uniformly bounded with respect to $k = 0, 1, \ldots$ for each $i \in Z$ and via (3.25) the sequence $\|u_k\|_i$ is uniformly bounded with respect to $k = 0, 1, \ldots$ for each $i \in Z$ too.

Let us prove now that the sequence $(x_k(i), u_k(i))$ is a fundamental sequence. From (3.9), (3.4) for each $i \in Z$, arbitrary $k = 0, 1, \ldots$ and $m = 0, 1, \ldots$ it follows that

$$\begin{aligned}&|x_{k+m}(i+1) - x_k(i+1)|\\ &\quad\le \sum_{j=0}^{i}|a(i,j)||x_{k+m}(j) - x_k(j)| + \sum_{j=0}^{i}|b(i,j)||u_{k+m}(j) - u_k(j)|\\ &\qquad + \varepsilon\left[\sum_{j=-h}^{i+1}|x_{k+m-1}(j) - x_{k-1}(j)|K_1(j) + \left|\sum_{j=0}^{i}(\sigma(i,j,x_{k+m-1}(j))\right.\right.\\ &\qquad\qquad \left.\left. - \sigma(i,j,x_{k-1}(j)))\xi(j+1)\right|\right].\end{aligned}$$

From this via (3.6) similarly to (3.22) we get

$$\mathbf{E}|x_{k+m}(i+1) - x_k(i+1)|^2 \leq C\left[\varepsilon^2\|x_{k+m-1} - x_{k-1}\|_N^2 + \sum_{j=0}^{i}\mathbf{E}|u_{k+m}(j) - u_k(j)|^2\right]. \tag{3.27}$$

Consider the second summand. For $j = 0$ from (3.10), (3.12) it follows that

$$\begin{aligned} u_{k+m}(0) - u_k(0) &= p(0)\left[\psi(N-1, 0, \mathbf{E}_0(\eta_{k+m}(\cdot+1) - \eta_k(\cdot+1)))\right] \\ &= p(0)\Bigg[\mathbf{E}_0(\eta_{k+m}(N) - \eta_k(N)) \\ &\qquad + \sum_{j=1}^{N-1} R(N-1, j)\mathbf{E}_0(\eta_{k+m}(j) - \eta_k(j))\Bigg] \\ &= \varepsilon p(0)\Bigg[\mathbf{E}_0(\Phi(N, x_{k+m-1,N}) - \Phi(N, x_{k-1,N})) \\ &\qquad + \sum_{j=1}^{N-1} R(N-1, j)\mathbf{E}_0(\Phi(j, x_{k+m-1,j}) \\ &\qquad - \Phi(j, x_{k-1,j}))\Bigg] \end{aligned}$$

From this via (3.4) and $x_{k+m-1}(l) = x_{k-1}(l)$ for $l \leq 0$ for some $C > 0$ we have

$$\begin{aligned} |u_{k+m}(0) - u_k(0)| &\leq \varepsilon C \sum_{j=1}^{N} \mathbf{E}_0|\Phi(j, x_{k+m-1,j}) - \Phi(j, x_{k-1,j})| \\ &\leq \varepsilon C \sum_{j=1}^{N}\sum_{l=1}^{j} \mathbf{E}_0|x_{k+m-1}(l) - x_{k-1}(l)|K_1(l) \\ &\leq \varepsilon C \sum_{l=1}^{N}\left(\sum_{j=l}^{N} K_1(l)\right)\mathbf{E}_0|x_{k+m-1}(l) - x_{k-1}(l)| \\ &\leq \varepsilon C K_1 N \sum_{l=1}^{N} \mathbf{E}_0|x_{k+m-1}(l) - x_{k-1}(l)|. \end{aligned}$$

Therefore,

$$\begin{aligned}\mathbf{E}|u_{k+m}(0)-u_k(0)|^2 &\le \varepsilon^2 C^2 K_1^2 N^2 \mathbf{E}\left(\sum_{j=1}^{N} \mathbf{E}_0|x_{k+m-1}(j)-x_{k-1}(j)|\right)^2 \\ &\le \varepsilon^2 C^2 K_1^2 N^3 \sum_{j=1}^{N} \mathbf{E}|x_{k+m-1}(j)-x_{k-1}(j)|^2\end{aligned}$$

and via (2.2) for $C_1 = C^2 K_1^2 N^4$ we obtain

$$\mathbf{E}|u_{k+m}(0)-u_k(0)|^2 \le \varepsilon^2 C_1 \|x_{k+m-1}-x_{k-1}\|_N^2. \tag{3.28}$$

For $j \ge 0$ from (3.10) we have

$$\begin{aligned}u_{k+m}(j+1)-u_k(j+1) &= \alpha_{k+m}(j+1)-\alpha_k(j+1) \\ &\quad + p(j+1)\psi(N-1,j,I)(x_{k+m}(j+1)-x_k(j+1)) \\ &\quad + \sum_{l=0}^{j}\gamma(j,l)(x_{k+m}(l)-x_k(l)) + Q(j,0)(u_{k+m}(0) \\ &\quad - u_k(0)).\end{aligned}$$

From this via (3.13), (3.14), and (3.28) similarly to (3.28) for some $C > 0$ and $j > 0$ it follows that

$$\begin{aligned}&\mathbf{E}|u_{k+m}(j)-u_k(j)|^2 \\ &\quad \le C\Big[\mathbf{E}|\alpha_{k+m}(j)-\alpha_k(j)|^2 + \|x_{k+m}-x_k\|_j^2 + \varepsilon^2\|x_{k+m-1}-x_{k-1}\|_N^2\Big].\end{aligned} \tag{3.29}$$

From (3.11), (3.13), (3.14), and (2.52) we obtain that for some $C > 0$

$$\begin{aligned}|\alpha_{k+m}(j+1)-\alpha_k(j+1)| &\le |p(j+1)\psi(N-1,j,\beta_{k+m}(\cdot,j+1) \\ &\quad - \beta_k(\cdot,j+1))| \\ &\quad + \sum_{l=1}^{j}|Q(j,l)p(l)\psi(N-1,l-1,\beta_{k+m}(\cdot,l) \\ &\quad - \beta_k(\cdot,l))| \\ &\le C\sum_{l=1}^{j+1}|\psi(N-1,l-1,\beta_{k+m}(\cdot,l)-\beta_k(\cdot,l))| \\ &\le C\sum_{l=1}^{j+1}\sum_{r=1}^{N-1}|\beta_{k+m}(r,l)-\beta_k(r,l)|.\end{aligned}$$

Via (3.12), (3.4) and $x_{k+m-1}(l) = x_{k-1}(l)$ for $l \leq 0$ we have

$$\begin{aligned}
|\beta_{k+m}(i,j) - \beta_k(i,j)| \leq \varepsilon \Bigg[& \sum_{l=1}^{i+1} \mathbf{E}_j |x_{k+m-1}(l) - x_{k-1}(l)| K_1(l) \\
& + \sum_{l=1}^{j} |x_{k+m-1}(l) - x_{k-1}(l)| K_1(l) \\
& + \left| \sum_{l=1}^{j-1} (\sigma(i,l,x_{k+m-1,l}) - \sigma(i,l,x_{k-1,l}))\xi(l+1) \right| \\
& + \Bigg| \sum_{l=1}^{j-1} (\sigma(j-1,l,x_{k+m-1,l}) \\
& - \sigma(j-1,l,x_{k-1,l}))\xi(l+1) \Bigg| \Bigg].
\end{aligned}$$

Squaring both parts of the obtained inequality, calculating the expectation and using (3.6), we obtain

$$\mathbf{E}|\beta_{k+m}(i,j) - \beta_k(i,j)|^2 \leq C\varepsilon^2 \|x_{k+m-1} - x_{k-1}\|_N^2.$$

So, it is easy to get that for some positive C the following estimate holds

$$\mathbf{E}|\alpha_{k+m}(j) - \alpha_k(j)|^2 \leq C\varepsilon^2 \|x_{k+m-1} - x_{k-1}\|_N^2.$$

As a result, via (3.28), (3.29) we obtain

$$\begin{aligned}
\mathbf{E}|u_{k+m}(j) - u_k(j)|^2 \leq C\left[\varepsilon^2 \|x_{k+m-1} - x_{k-1}\|_N^2 + \|x_{k+m} - x_k\|_j^2\right], \\
j = 0, 1, \ldots N-1. \qquad (3.30)
\end{aligned}$$

Substituting (3.30) into (3.27), we have

$$\begin{aligned}
& \mathbf{E}|x_{k+m}(i+1) - x_k(i+1)|^2 \\
& \leq C\left[\varepsilon^2 \|x_{k+m-1} - x_{k-1}\|_N^2 + \sum_{j=1}^{i} \|x_{k+m} - x_k\|_j^2\right].
\end{aligned}$$

Via Lemma 1.2 from this it follows that for some $C > 0$

$$\|x_{k+m} - x_k\|_{i+1}^2 \leq C\varepsilon^2 \|x_{k+m-1} - x_{k-1}\|_N^2.$$

Therefore,

$$\begin{aligned}\|x_{k+m} - x_k\|_{i+1}^2 &\leq (C\varepsilon^2)^2 \|x_{k+m-2} - x_{k-2}\|_N^2 \\ &\leq \dots \\ &\leq (C\varepsilon^2)^k \|x_m - x_0\|_N^2,\end{aligned} \tag{3.31}$$

i.e., for small enough ε such that $C\varepsilon^2 < 1$ for each $i \in Z$ and arbitrary $m = 0, 1, \dots$ we have

$$\lim_{k\to\infty} \|x_{k+m} - x_k\|_i = 0,$$

and via (3.30)

$$\lim_{k\to\infty} \|u_{k+m} - u_k\|_i = 0.$$

From the obtained properties of the sequences $(x_k(i), u_k(i))$ it follows [67, 184] that these sequences converge in mean square to some processes $(\tilde{x}(i), \tilde{u}(i))$. Let us show that these processes are connected by the conditions (3.15)–(3.18), (3.13), (3.14). Really, from (3.9) for $x = x_k$, $u = u_k$ and (3.11), (3.13)–(3.18) similarly to (3.31) for each $i \in Z$ it follows that

$$\|\tilde{x} - x_k\|_{i+1}^2 \leq (C\varepsilon^2)^k \|\tilde{x} - x_0\|_N^2. \tag{3.32}$$

So,

$$\lim_{k\to\infty} \|\tilde{x} - x_k\|_N = 0, \qquad \lim_{k\to\infty} \|\tilde{u} - u_k\|_N = 0.$$

The proof is completed.

Below it is shown that the controls u_k, $k = 0, 1, \dots,$ constructed above, are successive approximations to the optimal control of the control problem (3.1), (3.2), i.e., satisfy the condition (3.7).

3.3 A Zeroth Approximation

Consider the auxiliary control problem (3.8), (3.2). Optimal control of this control problem is known and is defined by (3.10)–(3.14) for $k = 0$.

Let $J(u)$ be the performance functional (3.2) for the Eq. (3.1) and $J_0(u)$ be the performance functional (3.2) for the linear equation (3.8). The closeness of the control problems (3.8), (3.2) and (3.1), (3.2) is defined by

$$\rho_0 = \sup_{u\in\mathbf{U}} |J(u) - J_0(u)|.$$

Really, putting $V_0 = \inf_{u \in \mathbf{U}} J_0(u) = J_0(u_0)$, via Lemma 1.1 and (1.11) we obtain

$$0 \leq J(u_0) - V \leq |J(u_0) - J_0(u_0)| + |V_0 - V| \leq 2\rho_0.$$

If $\rho_0 \leq C\varepsilon$ for some $C > 0$, then the control u_0 satisfies the condition (3.7) for $k = 0$, i.e., the control u_0 is a zeroth approximation to the optimal control of the control problem (3.1), (3.2).

Theorem 3.2 *The control $u_0(j)$ defined by* (3.10)–(3.14) *for $k = 0$ is a zeroth approximation to the optimal control of the control problem* (3.1), (3.2).

Proof It is enough to show that $\rho_0 \leq C\varepsilon$. Let x^u be a solution of the equation (3.1) and x_0^u be a solution of the equation (3.8) under the control $u = u(j)$. Then

$$\begin{aligned} |J(u) - J_0(u)| &= \mathbf{E}|(x^u(N))' F x^u(N) - (x_0^u(N))' F x_0^u(N)| \\ &= \mathbf{E}|(x^u(N) - x_0^u(N))' F (x^u(N) + x_0^u(N))| \\ &\leq F \sqrt{\mathbf{E}|x^u(N) - x_0^u(N)|^2 \mathbf{E}|x^u(N) + x_0^u(N)|^2} \\ &\leq F \|x^u - x_0^u\|_N \|x^u + x_0^u\|_N. \end{aligned} \tag{3.33}$$

Since $x^u \in H$ and $x_0^u \in H$ then it is enough to show that

$$\|x^u - x_0^u\|_N \leq C\varepsilon. \tag{3.34}$$

Subtracting (3.8) from (3.1) and using (3.3), we get

$$\begin{aligned} |x^u(i+1) - x_0^u(i+1)| \leq & \sum_{j=1}^{i} |a(i,j)||x^u(j) - x_0^u(j)| \\ & + \varepsilon \sum_{j=-h}^{i+1} (1 + |x^u(j)|) K_1(j) \\ & + \varepsilon \left| \sum_{j=0}^{i} \sigma(i, j, x_j^u) \xi(j+1) \right|. \end{aligned} \tag{3.35}$$

Squaring the both parts of the obtained inequality, calculating expectation and using (3.5), similarly to (3.25) we obtain

$$\mathbf{E}|x^u(i+1) - x_0^u(i+1)|^2 \leq C \left[\varepsilon^2 \left(1 + \|x^u\|_N^2\right) + \sum_{j=1}^{i} \mathbf{E}|x^u(j) - x_0^u(j)|^2 \right].$$

Put $y(i) = \mathbf{E}|x^u(i) - x_0^u(i)|^2$. Via Lemma 1.2 for some $C > 0$ we have $y(i+1) \leq C\varepsilon^2$, that is equivalent to (3.34). The proof is completed.

Now let us show that the control u_0 is a zeroth approximation to the optimal control of the control problem (3.1), (3.2) even though it is viewed not as a program control but as a feedback control. For this goal put

$$\begin{aligned}\hat{u}_0(0,x_0) &= p(0)\left[\psi(N-1,0,\mathbf{E}_0\eta_0(\cdot+1)) + R(N-1,0)\varphi_0(0)\right],\\ \hat{u}_0(j+1,x_{j+1}) &= \alpha_0(j+1) + p(j+1)\psi(N-1,j,I)x(j+1)\\ &\quad + \sum_{l=0}^{j}\gamma(j,l)x(l) + Q(j,0)\hat{u}_0(0,x_0), \quad j=0,1,\ldots,N-2.\end{aligned} \tag{3.36}$$

The process $\alpha_0(j)$ is defined by (3.11), (3.12), (3.13) for $k=0$, $\gamma(j,l)$ is defined by (3.14).

Theorem 3.3 *The control $\hat{u}_0(j,x_j)$ given by* (3.36) *is a zeroth approximation to the optimal control of the control problem* (3.1), (3.2), *i.e.,* $0 \le J(\hat{u}_0) - V \le C\varepsilon$.

Proof Note that the solution of the equation (3.8) under the control $\hat{u}_0$ coincides with the solution of this equation under the control u_0 given by (3.10)–(3.14) for $k=0$. So,

$$J_0(\hat{u}_0) = J_0(u_0) = V_0$$

and

$$0 \le J(\hat{u}_0) - V \le |J(\hat{u}_0) - J_0(\hat{u}_0)| + |V_0 - V|.$$

From the proof of Theorem 3.2 it follows that $|V_0 - V| \le C\varepsilon$, where C here and below denotes all positive constants. So, it is enough to show that

$$|J(\hat{u}_0) - J_0(\hat{u}_0)| \le C\varepsilon. \tag{3.37}$$

Let $\hat{x}_0$ and x_0 be the solutions of the equations (3.1) and (3.8) respectively under the control $\hat{u}_0$. From (3.2) similarly to (3.33) we get

$$\begin{aligned}&|J(\hat{u}_0) - J_0(\hat{u}_0)|\\ &\quad \le C\Bigg[\sqrt{\mathbf{E}|\hat{x}_0(N) - x_0(N)|^2\mathbf{E}|\hat{x}_0(N) + x_0(N)|^2}\\ &\qquad + \sum_{j=0}^{N-1}\sqrt{\mathbf{E}|\hat{u}_0(j,\hat{x}_{0j}) - \hat{u}_0(j,x_{0j})|^2\mathbf{E}|\hat{u}_0(j,\hat{x}_{0j}) + \hat{u}_0(j,x_{0j})|^2}\Bigg].\end{aligned} \tag{3.38}$$

Note that the processes $\hat{x}_0(j)$, $x_0(j)$, $\hat{u}_0(j,\hat{x}_{0j})$ and $\hat{u}_0(j,x_{0j})$ are processes from H and $\hat{x}_0(l) = x_0(l)$ for $l \le 0$. So,

$$|J(\hat{u}_0) - J_0(\hat{u}_0)| \le C\left[\|\hat{x}_0 - x_0\|_N + \sum_{j=1}^{N-1}\sqrt{\mathbf{E}|\hat{u}_0(j,\hat{x}_{0j}) - \hat{u}_0(j,x_{0j})|^2}\right],$$

and from (3.36), (3.11), (3.12) it follows that

$$
\begin{aligned}
&|\hat{u}_0(j+1, \hat{x}_{0,j+1}) - \hat{u}_0(j+1, x_{0,j+1})| \\
&\quad \leq |p(j+1)\psi(N-1, j, I)||\hat{x}_0(j+1) - x_0(j+1)| \\
&\quad + \sum_{l=1}^{j} |\gamma(j,l)||\hat{x}_0(l) - x_0(l)|.
\end{aligned}
$$

Squaring the both parts of the obtained inequality, calculating expectation and using (3.13), (3.14), we obtain

$$
\mathbf{E}|\hat{u}_0(j, \hat{x}_{0j}) - \hat{u}_0(j, x_{0j})|^2 \leq C\|\hat{x}_0 - x_0\|_j^2. \tag{3.39}
$$

As a result for some $C > 0$ we have

$$
|J(\hat{u}_0) - J_0(\hat{u}_0)| \leq C\|\hat{x}_0 - x_0\|_N. \tag{3.40}
$$

Let us estimate now $\|\hat{x}_0 - x_0\|_N$. Subtracting (3.8) from (3.1), by virtue of (3.3) we obtain

$$
\begin{aligned}
&|\hat{x}_0(i+1) - x_0(i+1)| \\
&\quad \leq \sum_{j=1}^{i} |a(i,j)||\hat{x}_0(j) - x_0(j)| + \sum_{j=0}^{i} |b(i,j)||\hat{u}_0(j, \hat{x}_{0j}) - \hat{u}_0(j, x_{0j})| \\
&\quad + \varepsilon\left[\sum_{j=0}^{i+1}\left(1 + |\hat{x}_{0,i+1}|\right)K_1(j) + \sum_{j=0}^{i} |\sigma(i,j,\hat{x}_{0j})\xi(j+1)|\right].
\end{aligned}
$$

From this via (3.5), (3.39) it follows that

$$
\mathbf{E}|\hat{x}_0(i+1) - x_0(i+1)|^2 \leq C\left[\varepsilon^2 + \sum_{j=1}^{i} \mathbf{E}|\hat{x}_0(j) - x_0(j)|^2\right].
$$

Via Lemma 1.2 it means that $\|\hat{x}_0 - x_0\|_N^2 \leq C\varepsilon^2$. From this and (3.40) the condition (3.37) follows. The proof is completed.

3.4 Approximations of Higher Orders

Consider the control problem (3.9), (3.2) as an auxiliary problem of control. The optimal control u_k of this control problem is defined by (3.10)–(3.14).

Let $J(u)$ be the performance functional (3.2) for the Eq. (3.1) and $J_k(u)$ be the performance functional (3.2) for the linear equation (3.9). The closeness of the control problems (3.9), (3.2) and (3.1), (3.2) is defined by

$$\rho_k = \sup_{u \in \mathbf{U}} |J(u) - J_k(u)|.$$

Really, put $V_k = \inf_{u \in \mathbf{U}} J_k(u) = J_k(u_k)$. Then via Lemma 1.1 and (1.11)

$$0 \le J(u_k) - V \le |J(u_k) - J_k(u_k)| + |V_k - V| \le 2\rho_k. \tag{3.41}$$

If $\rho_k \le C\varepsilon^{k+1}$, then the control u_k satisfies the condition (3.7). So, u_k is a kth approximation to the optimal control of the control problem (3.1), (3.2).

Theorem 3.4 *The control* $u_k(i)$, $k = 1, 2, \ldots$ *given by* (3.10)–(3.14) *is a kth approximation to the optimal control of the control problem* (3.1), (3.2).

Proof Let x^u and x_k^u, $k = 1, 2, \ldots$ be the solutions of the equations (3.1) and (3.8), respectively, under the control $u = u(j)$. Then similarly to (3.32), we have

$$\begin{aligned} |J(u) - J_k(u)| &= \mathbf{E}|(x^u(N) - x_k^u(N))' F (x^u(N) + x_k^u(N))| \\ &\le F\sqrt{\mathbf{E}|x^u(N) - x_k^u(N)|^2 \mathbf{E}|x^u(N) + x_k^u(N)|^2} \\ &\le F\|x^u - x_k^u\|_N \|x^u + x_k^u\|_N. \end{aligned}$$

So, it is enough to prove that

$$\|x^u - x_k^u\|_N \le C\varepsilon^{k+1}. \tag{3.42}$$

Subtracting (3.9) from (3.1) and using (3.4), we get

$$\begin{aligned} |x^u(i+1) - x_k^u(i+1)| \le{} & \sum_{j=1}^{i} |a(i,j)||x^u(j) - x_k^u(j)| \\ & + \varepsilon\Bigg[\sum_{j=1}^{i+1} |x^u(j) - x_{k-1}^u(j)| K_1(j) \\ & + \sum_{j=1}^{i} |\sigma(i,j,x_j^u) - \sigma(i,j,x_{k-1,j}^u)| |\xi(j+1)|\Bigg]. \end{aligned}$$

Squaring the both parts of the obtained inequality, calculating expectation and using (3.6), we obtain for some $C > 0$

$$\mathbf{E}|x^u(i+1) - x_k^u(i+1)|^2 \le C\left[\varepsilon^2\|x^u - x_{k-1}^u\|_N^2 + \sum_{j=1}^{i}\mathbf{E}|x^u(j) - x_k^u(j)|^2\right].$$

From this and Lemma 1.2 similarly to the previous proofs it follows that for some $C > 0$

$$\|x^u - x_k^u\|_N \le C\varepsilon\|x^u - x_{k-1}^u\|_N.$$

So,

$$\begin{aligned}\|x^u - x_k^u\|_N &\le C\varepsilon\|x^u - x_{k-1}^u\|_N \\ &\le (C\varepsilon)^2\|x^u - x_{k-2}^u\|_N \\ &\le \dots \\ &\le (C\varepsilon)^k\|x^u - x_0^u\|_N.\end{aligned}$$

From this and (3.34) the condition (3.42) follows. Theorem is proven.

Let us show that the control u_k, $k = 1, 2, \dots$, is a kth approximation to the optimal control of the control problem (3.1), (3.2) even though it is considered not as a program control but as a feedback control. Rewrite (3.10) in the form

$$\begin{aligned}\hat{u}_k(0, x_0) &= p(0)\left[\psi(N-1, 0, \mathbf{E}_0\eta_k(\cdot + 1)) + R(N-1, 0)\varphi_0(0)\right], \\ \hat{u}_k(j+1, x_{j+1}) &= \alpha_k(j+1) + p(j+1)\psi(N-1, j, I)x(j+1) \\ &\quad + \sum_{l=0}^{j}\gamma(j, l)x(l) + Q(j, 0)\hat{u}_k(0, x_0), \quad j = 0, 1, \dots, N-2,\end{aligned} \tag{3.43}$$

where $\alpha_k(j)$ and $\gamma(j, l)$ are defined by (3.11)–(3.14).

Theorem 3.5 *The control $\hat{u}_k(j, x_j)$, $k = 1, 2, \dots$, given by* (3.43) *is a kth approximation to the optimal control of the control problem* (3.1), (3.2), *i.e.*,

$$0 \le J(\hat{u}_k) - V \le C\varepsilon^{k+1}.$$

Proof Note that the solution of the equation (3.9) under the control $\hat{u}_k$ coincides with the solution of this equation under the control u_k given by (3.10)–(3.14). So,

$$J_k(\hat{u}_k) = J_k(u_k) = V_k$$

and

$$0 \le J(\hat{u}_k) - V \le |J(\hat{u}_k) - J_k(\hat{u}_k)| + |V_k - V|.$$

From (3.41) and the proof of Theorem 3.4 it follows that $|V_k - V| \le C\varepsilon^{k+1}$, where C here and below denotes all positive constants. So, it is enough to show that

$$|J(\hat{u}_k) - J_k(\hat{u}_k)| \le C\varepsilon^{k+1}. \tag{3.44}$$

Let $\hat{x}_k$ and x_k be the solutions of the equations (3.1) and (3.9), respectively, under the control $\hat{u}_k$. From (3.2) similarly to (3.38), we get

$$|J(\hat{u}_k) - J_k(\hat{u}_k)| \le C\Bigg[\sqrt{\mathbf{E}|\hat{x}_k(N) - x_k(N)|^2\mathbf{E}|\hat{x}_k(N) + x_k(N)|^2}$$
$$+ \sum_{j=0}^{N-1}\sqrt{\mathbf{E}|\hat{u}_k(j,\hat{x}_{kj}) - \hat{u}_k(j,x_{kj})|^2\mathbf{E}|\hat{u}_k(j,\hat{x}_{kj}) + \hat{u}_k(j,x_{kj})|^2}\Bigg].$$

Since the processes $\hat{x}_k(j)$, $x_k(j)$, $\hat{u}_k(j,\hat{x}_{kj})$, and $\hat{u}_k(j,x_{kj})$ are members of H and $\hat{x}_k(l) = x_k(l)$ for $l \le 0$, then

$$|J(\hat{u}_k) - J_k(\hat{u}_k)| \le C\Bigg[\|\hat{x}_k - x_k\|_N + \sum_{j=1}^{N-1}\sqrt{\mathbf{E}|\hat{u}_k(j,\hat{x}_{kj}) - \hat{u}_k(j,x_{kj})|^2}\Bigg]. \tag{3.45}$$

From (3.43), we have

$$|\hat{u}_k(j+1,\hat{x}_{k,j+1}) - \hat{u}_k(j+1,x_{k,j+1})| \le |p(j+1)\psi(N-1,j,I)||\hat{x}_k(j+1)$$
$$- x_k(j+1)| + \sum_{l=1}^{j}|\gamma(j,l)||\hat{x}_k(l) - x_k(l)|.$$

From this via (3.13), (3.14) we have

$$\mathbf{E}|\hat{u}_k(j,\hat{x}_{kj}) - \hat{u}_k(j,x_{kj})|^2 \le C\|\hat{x}_k - x_k\|_j^2, \quad j = 1,\dots,N-1, \tag{3.46}$$

and from (3.45), (3.46) it follows that

$$|J(\hat{u}_k) - J_k(\hat{u}_k)| \le C\|\hat{x}_k - x_k\|_N. \tag{3.47}$$

Let $\tilde{x}$ be the process defined by (3.15), (3.16). It is clear that

$$\|\hat{x}_k - x_k\|_N \le C(\|\hat{x}_k - \tilde{x}\|_N + \|\tilde{x} - x_k\|_N). \tag{3.48}$$

Similarly to (3.34) one can show that $\|\tilde{x} - x_0\|_N \le C\varepsilon$. From this and (3.32), we obtain

$$\|\tilde{x} - x_k\|_N \le C\varepsilon^{k+1}. \tag{3.49}$$

So, via (3.47)–(3.49) it is enough to prove that

$$\|\hat{x}_k - \tilde{x}\|_N \le C\varepsilon^{k+1}. \tag{3.50}$$

Substituting the control (3.43) into the Eq. (3.1), we get the following equation for the process $\hat{x}_k$:

$$\begin{aligned}
\hat{x}_k(i+1) &= \eta(i+1) + \sum_{j=0}^{i} a(i,j)\hat{x}_k(j) \\
&\quad + b(i,0)p(0)\left[\psi(N-1,0,\mathbf{E}_0\eta_k(\cdot+1)) + R(N-1,0)\varphi_0(0)\right] \\
&\quad + \sum_{j=1}^{i} b(i,j)\Bigg[\alpha_k(j) + p(j)\psi(N-1,j-1,I)\hat{x}_k(j) \\
&\qquad\qquad + \sum_{l=0}^{j-1}\gamma(j-1,l)\hat{x}_k(l) + Q(j,0)\hat{u}_k(0,x_0)\Bigg] \\
&\quad + \varepsilon\Bigg[\Phi(i+1,\hat{x}_{k,i+1}) + \sum_{j=0}^{i}\sigma(i,j,\hat{x}_{kj})\xi(j+1)\Bigg].
\end{aligned} \tag{3.51}$$

Substituting (3.16) into (3.15), we obtain the equation for the process $\tilde{x}(i)$:

$$\begin{aligned}
\tilde{x}(i+1) &= \eta(i+1) + \sum_{j=0}^{i} a(i,j)\tilde{x}(j) \\
&\quad + b(i,0)p(0)\left[\psi(N-1,0,\mathbf{E}_0\tilde{\eta}(\cdot+1)) + R(N-1,0)\varphi_0(0)\right] \\
&\quad + \sum_{j=1}^{i} b(i,j)\Bigg[\alpha(j) + p(j)\psi(N-1,j-1,I)\tilde{x}(j) \\
&\qquad\qquad + \sum_{l=0}^{j-1}\gamma(j-1,l)\tilde{x}(l) + Q(j,0)\hat{u}_k(0)\Bigg] \\
&\quad + \varepsilon\Bigg[\Phi(i+1,\tilde{x}_{i+1}) + \sum_{j=0}^{i}\sigma(i,j,\tilde{x}_j)\xi(j+1)\Bigg].
\end{aligned} \tag{3.52}$$

Subtracting (3.52) from (3.51) and using the conditions $\hat{u}_k(0) = \hat{u}_k(0, x_0)$ and (3.4), we have

$$
\begin{aligned}
&|\hat{x}_k(i+1) - \tilde{x}(i+1)| \\
&\quad \le \sum_{j=1}^{i} |a(i,j)||\hat{x}_k(j) - \tilde{x}(j)| \\
&\quad + |b(i,0)p(0)||\psi(N-1, 0, \mathbf{E}_0(\eta_k(\cdot+1) - \tilde{\eta}(\cdot+1)))| \\
&\quad + \sum_{j=1}^{i} |b(i,j)| \Big[|\alpha_k(j) - \alpha(j)| + |p(j)\psi(N-1, j-1, I)||\hat{x}_k(j) \\
&\qquad\qquad - \tilde{x}(j)| + \sum_{l=1}^{j-1} |\gamma(j-1,l)||\hat{x}_k(l) - \tilde{x}(l)| \Big] \\
&\quad + \varepsilon \Big[\sum_{j=1}^{i+1} |\hat{x}_k(j) - \tilde{x}(j)| K_1(j) + \sum_{j=1}^{i} |(\sigma(i,j,\hat{x}_{kj}) - \sigma(i,j,\tilde{x}_j))\xi(j+1)| \Big].
\end{aligned}
\tag{3.53}
$$

Squaring the both parts of the inequality (3.53), calculating expectation and using (2.52), (3.6), we obtain

$$
\begin{aligned}
\mathbf{E}|\hat{x}_k(i+1) - \tilde{x}(i+1)|^2 \le C \Big[&\varepsilon^2 \|\hat{x}_k - \tilde{x}\|_N^2 + \sum_{j=1}^{N} \mathbf{E}|\eta_k(j) - \tilde{\eta}(j)|^2 \\
&+ \sum_{j=0}^{i} \mathbf{E}|\alpha_k(j) - \alpha(j)|^2 + \sum_{j=1}^{i} \mathbf{E}|\hat{x}_k(j) - \tilde{x}(j)|^2 \Big].
\end{aligned}
\tag{3.54}
$$

Using (3.12), (3.18), (3.4), and (3.49), we have

$$
\begin{aligned}
\mathbf{E}|\eta_k(j) - \tilde{\eta}(j)|^2 &\le \varepsilon^2 \mathbf{E}|\Phi(j, x_{k-1,j}) - \Phi(j, \tilde{x}_j)|^2 \\
&\le C\varepsilon^2 \|x_{k-1} - \tilde{x}\|^2 \\
&\le C\varepsilon^{2(k+1)}.
\end{aligned}
\tag{3.55}
$$

From (3.11), (3.17) and (2.52) it follows that

$$
\begin{aligned}
\alpha_k(j) - \alpha(j) = &\; p(j)[\psi(N-1, j-1, \beta_k(\cdot, j) - \beta(\cdot, j))] \\
&+ \sum_{l=0}^{j-1} Q(j-1, l)p(l)[\psi(N-1, l-1, \beta_k(\cdot, l) - \beta(\cdot, l))].
\end{aligned}
\tag{3.56}
$$

Note that via (2.52)

$$\psi(N-1, j-1, \beta_k(\cdot, j) - \beta(\cdot, j))$$
$$= \beta_k(N-1, j) - \beta(N-1, j) + \sum_{l=j}^{N-1} R(N-1, l)[\beta_k(l-1, j) - \beta(l-1, j)].$$

So,

$$\mathbf{E}|\psi(N-1, j-1, \beta_k(\cdot, j) - \beta(\cdot, j))|^2 \leq C \max_{j-1 \leq l \leq N-1} \mathbf{E}|\beta_k(l, j) - \beta(l, j)|^2. \tag{3.57}$$

From (3.12), (3.18) it follows that

$$\beta_k(i, j) - \beta_0(i, j) = \varepsilon \Bigg[\mathbf{E}_j(\Phi(i+1, x_{k-1,i+1}) - \Phi(i+1, \tilde{x}_{i+1})) + \Phi(j, \tilde{x}_j)$$
$$- \Phi(j, x_{k-1,j}) + \sum_{l=1}^{j-1} \Big[\sigma(i, l, x_{k-1,l}) - \sigma(i, l, \tilde{x}_l) \Big] \xi(l+1)$$
$$+ \sum_{l=1}^{j-1} \Big[\sigma(j-1, l, \tilde{x}_l) - \sigma(j-1, l, x_{k-1,l}) \Big] \xi(l+1) \Bigg].$$

From this via (3.4), (3.6), and (3.49), we have

$$\mathbf{E}|\beta_k(i, j) - \beta(i, j)|^2 \leq C\varepsilon^2 \|x_{k-1} - \tilde{x}\|_j^2$$
$$\leq C\varepsilon^{2(k+1)}.$$

From this and (3.57), (5.56) it follows that

$$\mathbf{E}|\alpha_k(j) - \alpha(j)|^2 \leq C\varepsilon^{2(k+1)}. \tag{3.58}$$

As a result from (3.54), (3.55), (3.58) we obtain

$$\mathbf{E}|\hat{x}_k(i+1) - \tilde{x}(i+1)|^2 \leq C\Bigg[\varepsilon^2 \|\hat{x}_k - \tilde{x}\|_N^2 + \varepsilon^{2(k+1)} + \sum_{j=1}^{i} \mathbf{E}|\hat{x}_k(j) - \tilde{x}(j)|^2 \Bigg].$$

Using Lemma 1.2, for some $C > 0$ we have

$$\mathbf{E}|\hat{x}_k(i+1) - \tilde{x}(i+1)|^2 \leq C\Big[\varepsilon^2 \|\hat{x}_k - \tilde{x}\|_N^2 + \varepsilon^{2(k+1)} \Big].$$

From this for small enough ε (such that $C\varepsilon^2 < 1$) the condition (3.50) follows. The proof is completed.

Chapter 4
Optimal and Quasioptimal Stabilization

Here the problem of the optimal stabilization for a linear stochastic difference Volterra equation and quadratic performance functional is considered. Optimal control in the sense of a given quadratic performance functional that stabilizes the solution of the considered equation to mean square stable and mean square summable is constructed. For a quasilinear stochastic difference Volterra equation with quadratic performance functional a zero approximation to the optimal control constructed that stabilizes the solution of the considered equation to mean square stable and mean square summable is constructed.

4.1 Statement of the Linear Quadratic Optimal Stabilization Problem

Consider the problem of optimal control for the linear stochastic difference Volterra equation

$$x(i+1) = \eta(i+1) + \sum_{j=0}^{i} a(i,j)x(j) + \sum_{j=0}^{i} b(i,j)u(j),$$
$$i \geq 0, \qquad x(0) = \varphi_0(0), \tag{4.1}$$

and the quadratic performance functional of the form

$$J(u) = \mathbf{E}\sum_{j=0}^{\infty}\left[x'(j)F(j)x(j) + u'(j)G(j)u(j)\right]. \tag{4.2}$$

Here $\eta \in H$, $a(i,j)$ and $b(i,j)$ are nonrandom matrices of the $n \times n$ and $n \times m$ dimensions respectively, $u(j) \in \mathbf{R}^m$. It is supposed also that the positive semidefinite nonrandom matrix $F(j)$ and positive definite nonrandom matrix $G(j)$ of the

L. Shaikhet, *Optimal Control of Stochastic Difference Volterra Equations*,
Studies in Systems, Decision and Control 17, DOI 10.1007/978-3-319-13239-6_4

dimensions $n \times n$ and $m \times m$ respectively are uniformly bounded over to $j \geq 0$, the matrix $G(j)$ has the inverse matrix such that $\sup_{j\geq 0} |G^{-1}(j)| < \infty$.

Definition 4.1 The solution of the stochastic difference equation

$$x(i+1) = \eta(i+1) + A(i, x(0), x(1), \dots, x(i)),$$
$$i \geq 0, \qquad x(0) = \eta(0),$$

is called mean square stable if for every $\varepsilon > 0$ there exists a $\delta > 0$ such that

$$\|x\|^2 = \sup_{i\geq 0} \mathbf{E}|x(i)|^2 < \varepsilon \quad \text{if} \quad \|\eta\|^2 = \sup_{i\geq 0} \mathbf{E}|\eta(i)|^2 < \delta.$$

Definition 4.2 The sequence $x(i)$, $i \geq 0$, is called:

- uniformly mean square bounded if $\|x\|^2 = \sup_{i\geq 0} \mathbf{E}|x(i)|^2 < \infty$;
- asymptotically mean square trivial if $\lim_{i\to\infty} \mathbf{E}\left|x(i)\right|^2 = 0$;
- mean square summable if $S(x) = \sum_{i=0}^{\infty} \mathbf{E}|x(i)|^2 < \infty$.

Remark 4.1 Note that if the sequence $x(i)$, $i \geq 0$, is mean square summable then it is uniformly mean square bounded and asymptotically mean square trivial.

It is necessary to find such admissible control u_0, for which the solution of the equation (4.1) is mean square stable, mean square summable and for which the functional (4.2) be minimized, i.e., $J(u_0) = \inf_{u\in U} J(u)$.

The considered problem is solved in three steps. At first step, sufficient conditions for mean square stability and mean square summability of the solution of the linear stochastic difference Volterra equation

$$x(i+1) = \eta(i+1) + \sum_{j=0}^{i} a(i,j)x(j),$$
$$i \geq 0, \qquad x(0) = \varphi_0(0), \tag{4.3}$$

with stochastic perturbations $\eta(i)$ are obtained. At second step, the optimal control $u_0(j)$ of the control problem (4.1) and (4.2) is constructed. At third step, sufficient conditions for mean square stability and mean square summability are obtained for the solution of the difference equation (4.1) under the optimal control $u_0(j)$.

4.1.1 Auxiliary Stability Problem

Let $R(i,j)$ be the resolvent of the kernel $a(i,j)$ of the difference equation (4.3) and $|R(i,j)|$ be the operator norm of a matrix $R(i,j)$. Put also $\eta(0) = x(0) = \varphi_0(0)$.

Lemma 4.1 *If*

$$R = \sup_{i\geq 0}\sum_{j=0}^{i}|R(i,j)| < \infty \tag{4.4}$$

then the solution of (4.3) *is mean square stable. If besides*

$$\hat{R} = \sup_{j\geq 0}\sum_{i=j}^{\infty}|R(i,j)| < \infty \tag{4.5}$$

and the sequence $\eta(i)$, $i = 0, 1, \ldots$, is mean square summable then the solution of the equation (4.3) *is mean square summable too.*

Proof Using the resolvent $R(i,j)$ of the kernel $a(i,j)$ of the Eq. (4.3), via (1.17) we have

$$x(i+1) = \eta(i+1) + \sum_{j=0}^{i} R(i,j)\eta(j). \tag{4.6}$$

From this via Lemma 1.8 for arbitrary $\alpha > 0$ it follows that

$$\mathbf{E}|x(i+1)|^2 \leq (1+\alpha)\mathbf{E}|\eta(i+1)|^2 + \left(1+\frac{1}{\alpha}\right)\mathbf{E}\left|\sum_{j=0}^{i} R(i,j)\eta(j)\right|^2.$$

Via (4.4) we obtain

$$\begin{aligned}\left|\sum_{j=0}^{i} R(i,j)\eta(j)\right|^2 &\leq \sum_{l=0}^{i}|R(i,l)|\sum_{j=0}^{i}|R(i,j)||\eta(j)|^2\\ &\leq R\sum_{j=0}^{i}|R(i,j)||\eta(j)|^2.\end{aligned}$$

So,

$$\mathbf{E}|x(i+1)|^2 \leq (1+\alpha)\mathbf{E}|\eta(i+1)|^2 + \left(1+\frac{1}{\alpha}\right)R\sum_{j=0}^{i}|R(i,j)|\mathbf{E}|\eta(j)|^2. \tag{4.7}$$

From this for $\alpha = R$ by virtue of (4.4) it follows that $\|x\|^2 \leq (1+R)^2\|\eta\|^2$. Therefore, for arbitrary $\varepsilon > 0$ we have $\|x\|^2 < \varepsilon$ if $\|\eta\|^2 < \delta = \varepsilon(1+R)^{-2}$. So, mean square stability is proven.

Summing the inequality (4.7) over $i \geq 0$ together with the trivial inequality

$$\mathbf{E}|x(0)|^2 \leq (1+\alpha)\mathbf{E}|\eta(0)|^2,$$

we obtain

$$\sum_{i=0}^{\infty} \mathbf{E}|x(i)|^2 \leq (1+\alpha)\sum_{i=0}^{\infty} \mathbf{E}|\eta(i)|^2 + \left(1+\frac{1}{\alpha}\right)R\sum_{i=0}^{\infty}\sum_{j=0}^{i}|R(i,j)|\mathbf{E}|\eta(j)|^2.$$

Changing the order of summation, via (4.5) and $\alpha = \sqrt{R\hat{R}}$ for

$$S(x) = \sum_{i=0}^{\infty} \mathbf{E}|x(i)|^2$$

we get

$$\begin{aligned} S(x) &\leq (1+\alpha)S(\eta) + \left(1+\frac{1}{\alpha}\right)R\sum_{j=0}^{\infty}\sum_{i=j}^{\infty}|R(i,j)|\mathbf{E}|\eta(j)|^2 \\ &\leq \left[1+\alpha+\left(1+\frac{1}{\alpha}\right)R\hat{R}\right]S(\eta) \\ &= \left(1+\sqrt{R\hat{R}}\right)^2 S(\eta). \end{aligned}$$

So, if $\eta(j)$ is mean square summable, i.e., $S(\eta) < \infty$, then the solution of the equation (4.3) is mean square summable too, i.e., $S(x) < \infty$. The proof is completed.

Lemma 4.2 *If*

$$A = \sup_{i\geq 0}\sum_{j=0}^{i}|a(i,j)| < 1 \tag{4.8}$$

then the solution of the equation (4.3) *is mean square stable. If besides*

$$\hat{A} = \sup_{j\geq 0}\sum_{i=j}^{\infty}|a(i,j)| < 1 \tag{4.9}$$

and the sequence $\eta(i)$, $i = 0, 1, \ldots$, *is mean square summable then the solution of the equation* (4.3) *is mean square summable too.*

Proof Via Lemma 1.3 and (1.19) the kernel $a(i,j)$ and its resolvent $R(i,j)$ are connected by the equation

$$R(i,j) = a(i,j) + \sum_{l=j+1}^{i} R(i,l)a(l-1,j), \qquad 0 \le j \le i. \tag{4.10}$$

Summing (4.10) over $j = 0, \dots, i$ and changing the order of summation, by virtue of (4.8) we obtain

$$\begin{aligned}
\sum_{j=0}^{i} |R(i,j)| &\le \sum_{j=0}^{i} |a(i,j)| + \sum_{j=0}^{i} \sum_{l=j+1}^{i} |R(i,l)||a(l-1,j)| \\
&\le \sum_{j=0}^{i} |a(i,j)| + \sum_{l=1}^{i} |R(i,l)| \sum_{j=0}^{l-1} |a(l-1,j)| \\
&\le A + A \sum_{l=0}^{i} |R(i,l)|.
\end{aligned}$$

From this and (4.8) we get

$$\sum_{l=0}^{i} |R(i,l)| \le \frac{A}{1-A} < \infty. \tag{4.11}$$

So, the condition (4.4) holds. From Lemma 4.1, it follows that the solution of the equation (4.3) is mean square stable.

Let now $j \le N$. Summing (4.10) over $i = j, \dots, N$ and changing the order of summation, by virtue of (4.9) we have

$$\begin{aligned}
\sum_{i=j}^{N} |R(i,j)| &\le \sum_{i=j}^{N} |a(i,j)| + \sum_{i=j}^{N} \sum_{l=j+1}^{i} |R(i,l)||a(l-1,j)| \\
&\le \sum_{i=j}^{N} |a(i,j)| + \sum_{l=j+1}^{N} \sum_{i=l}^{N} |R(i,l)||a(l-1,j)| \\
&\le \hat{A} + \hat{A} \sup_{0 \le l \le N} \sum_{i=l}^{N} |R(i,l)|.
\end{aligned}$$

From this and (4.9) it follows that

$$\sup_{0\leq l\leq N}\sum_{i=l}^{N}|R(i,l)|\leq\frac{\hat{A}}{1-\hat{A}}<\infty.$$

Calculating the limit over $N\to\infty$, we obtain (4.5). From Lemma 4.1, it follows that the solution of the equation (4.3) is mean square summable. Lemma is proven.

Remark 4.2 Consider the scalar difference equation

$$x(i+1)=\eta(i+1)+\sum_{j=0}^{i}a(i-j)x(j)$$

or

$$x(i+1)=\eta(i+1)+a(i)x(0)+a(i-1)x(1)+\cdots+a(0)x(i).$$

In this case $a(i,j)=a(i-j)$ and via (4.8), (4.9)

$$A=\sup_{i\geq 0}\sum_{j=0}^{i}|a(i-j)|=\sum_{i=0}^{\infty}|a(i)|,$$

$$\hat{A}=\sup_{j\geq 0}\sum_{i=j}^{\infty}|a(i-j)|=\sum_{i=0}^{\infty}|a(i)|,$$

i.e., the conditions (4.8) and (4.9) coincide and have the form

$$A=\hat{A}=\sum_{i=0}^{\infty}|a(i)|<1.$$

So, via Lemma 4.2 if $A<1$ and the sequence $\eta(i)$, $i=0,1,\ldots$, is mean square summable then the solution of the considered equation is mean square stable and mean square summable.

Similarly, for $R(i,j)=R(i-j)$ the conditions (4.4) and (4.5) coincide too and have the form

$$R=\hat{R}=\sum_{i=0}^{\infty}|R(i)|<\infty.$$

Remark 4.3 Let us show that the solution of some difference equation can be mean square stable but not mean square summable. Consider the scalar difference equation

$$x(i+1)=\eta(i+1)+\sum_{j=0}^{i}a(j)x(j)$$

or

$$x(i+1) = \eta(i+1) + a(0)x(0) + a(1)x(1) + \cdots + a(i)x(i).$$

In this case $a(i, j) = a(j)$, the condition (4.8) takes the form

$$A = \sup_{i \geq 0} \sum_{j=0}^{i} |a(j)| = \sum_{j=0}^{\infty} |a(j)| < 1,$$

but the condition (4.9) does not hold (if there exists at least one nonzero $a(j)$)

$$\hat{A} = \sup_{j \geq 0} \sum_{i=j}^{\infty} |a(j)| = \infty.$$

So, the solution of the considered equation is mean square stable but can be not mean square summable.

Remark 4.4 From (4.6), it follows that the condition

$$\lim_{i \to \infty} \mathbf{E}|x(i)|^2 = 0$$

is equivalent to

$$\lim_{i \to \infty} \mathbf{E} \left| \eta(i+1) + \sum_{j=0}^{i} R(i, j)\eta(j) \right|^2 = 0. \tag{4.12}$$

Therefore, by the condition (4.12) the solution of the equation (4.3) is asymptotically mean square trivial.

Note that for asymptotic mean square triviality of the solution of the equation (4.3) it is not necessary to suppose asymptotic mean square triviality of the sequence $\eta(i)$, $i \in Z$. Let be, for example, $\eta(i) = x(0)$, $i \geq 0$, $\mathbf{E}|x(0)|^2 < \infty$. Then via (4.6)

$$x(i+1) = \left(I + \sum_{j=0}^{i} R(i, j) \right) x(0)$$

and via the condition $\lim_{i \to \infty} \left| I + \sum_{j=0}^{i} R(i, j) \right| = 0$ the solution of the equation (4.3) is asymptotically mean square trivial.

Example 4.1 Consider the scalar difference Volterra equation

$$x(i+1) = \eta(i+1) + b\sum_{j=0}^{i} a^{i-j}x(j),$$
$$x(0) = \varphi_0(0), \qquad i = 0, 1, \ldots. \tag{4.13}$$

Using Remark 4.2 and the condition $|a| < 1$, we have

$$A = \hat{A} = |b|\sum_{i=0}^{\infty} |a|^i = \frac{|b|}{1-|a|} < 1. \tag{4.14}$$

From Lemma 4.2, it follows that if

$$|a| + |b| < 1 \tag{4.15}$$

then the solution of the equation (4.13) is mean square stable. If besides the sequence $\eta(i)$ is mean square summable then the solution of the equation (4.13) is mean square summable too.

Let us calculate the resolvent of the kernel $a(i) = ba^i$ of the Eq. (4.13) and show that by the condition (4.15) the resolvent satisfies the condition (4.11).

From (1.21) we have

$$\begin{aligned} R(0) &= b, \\ R(1) &= ba + R(0)b \\ &= ba + b^2 \\ &= b(a+b), \\ R(2) &= ba^2 + R(1)b + R(0)ba \\ &= b(a^2 + b(a+b) + ba) \\ &= b(a+b)^2, \end{aligned}$$

....

Let us check that the resolvent $R(i)$ of the kernel $a(i) = ba^i$ of the Eq. (4.13) equals

$$R(i) = b(a+b)^i, \qquad i = 0, 1, \ldots. \tag{4.16}$$

Really, substituting (4.16) into the recurrent formula (1.20) defined a resolvent, we obtain

$$b(a+b)^i = ba^i + \sum_{j=0}^{i-1} ba^{i-1-j}b(a+b)^j.$$

If $a = 0$ then this equality holds trivially. If $a \neq 0$ then from this we have

$$(1+q)^i = 1 + q\sum_{j=0}^{i-1}(1+q)^j, \qquad q = \frac{b}{a}. \tag{4.17}$$

Using binomial theorem from (4.17) we obtain

$$\begin{aligned}
1 + \sum_{m=1}^{i}\binom{i}{m}q^m &= 1 + q\sum_{j=0}^{i-1}\sum_{k=0}^{j}\binom{j}{k}q^k \\
&= 1 + \sum_{k=0}^{i-1}\sum_{j=k}^{i-1}\binom{j}{k}q^{k+1} \\
&= 1 + \sum_{m=1}^{i}\sum_{j=m-1}^{i-1}\binom{j}{m-1}q^m
\end{aligned}$$

or known formula (1.34) for a number of combinations

$$\binom{i}{m} = \sum_{j=m-1}^{i-1}\binom{j}{m-1}, \qquad m = 1, \ldots, i.$$

So, (4.16) is the resolvent of the kernel $a(i) = ba^i$ of the Eq. (4.13). Substituting (4.16) into (4.11) and using (4.15), (4.14), we obtain that the condition (4.11) holds

$$\begin{aligned}
\sum_{l=0}^{i}|b(a+b)^{i-l}| &\le |b|\sum_{l=0}^{\infty}|a+b|^l \\
&= \frac{|b|}{1-|a+b|} \\
&\le \frac{|b|}{1-|a|-|b|} \\
&= \frac{|b|(1-|a|)^{-1}}{1-|b|(1-|a|)^{-1}} \\
&= \frac{A}{1-A} < \infty.
\end{aligned}$$

4.1.2 Optimal Control Problem

Consider the optimal control problem (4.1), (4.2) and construct the synthesis of optimal control for this problem. To realize this goal, we have to get the following auxiliary statements.

4.1.2.1 Auxiliary Representation of Optimal Control

Lemma 4.3 *The optimal control u_0 of the control problem* (4.1), (4.2) *has the representation*

$$u_0(j) = \sum_{i=j}^{\infty} \gamma(i,j)\mathbf{E}_j x_0(i+1), \qquad j \geq 0, \tag{4.18}$$

where

$$\gamma(i,j) = -G^{-1}(j)\psi'(i,j,b(\cdot,j))F(i+1), \qquad i \geq j \geq 0, \tag{4.19}$$

and $\psi(i,j,b(\cdot,j))$ is defined in (2.52).

Proof Let x_0 and x_ε be the solutions of the equation (4.1) by the controls u_0 and $u_\varepsilon = u_0 + \varepsilon v$ respectively. From (4.2) it follows that

$$\begin{aligned}
\frac{1}{\varepsilon}[J(u_\varepsilon) - J(u_0)] &= \frac{1}{\varepsilon}\sum_{j=0}^{\infty} \mathbf{E}[x_\varepsilon'(j)F(j)x_\varepsilon(j) - x_0'(j)F(j)x_0(j) \\
&\quad + (u_0'(j) + \varepsilon v(j))'G(j)(u_0(j) + \varepsilon v(j)) \\
&\quad - u_0'(j)G(j)u_0(j)] \\
&= \sum_{j=0}^{\infty} \mathbf{E}[q_\varepsilon'(j)F(j)x_\varepsilon(j) + x_0'(j)F(j)q_\varepsilon(j) \\
&\quad + 2u_0'(j)G(j)v(j) + \varepsilon v'(j)G(j)v(j)],
\end{aligned} \tag{4.20}$$

where $q_\varepsilon(j) = \frac{1}{\varepsilon}[x_\varepsilon(j) - x_0(j)]$.

From (4.1) it follows that $x_\varepsilon(j)$ is linear with respect to ε, $q_\varepsilon(j)$ does not depend on ε, i.e., $q_\varepsilon(i) = q_0(i)$ and $q_0(i)$ is defined by (2.54), i.e.,

$$q_0(i+1) = \sum_{j=0}^{i} \psi(i,j,b(\cdot,j))v(j), \quad i \in Z, \quad q_0(0) = 0. \tag{4.21}$$

Note that u_0 and u_ε are admissible controls. So, for small enough $\varepsilon_0 > 0$ the series in (4.20) converge uniformly with respect to $\varepsilon \in [0, \varepsilon_0]$ and the limit under sign of sum is possible [184]. Since $q_\varepsilon(j) = q_0(j)$ and $x_\varepsilon(j) \to x_0(j)$ then calculating in (4.20) the limit with respect to $\varepsilon \to 0$, we get

$$J_0(u_0) = 2\mathbf{E}\sum_{j=0}^{\infty}[x_0'(j)F(j)q_0(j) + u_0'(j)G(j)v(j)]. \tag{4.22}$$

Substituting (4.21) into (4.22) and using that $q_0(0) = 0$, we obtain

$$
\begin{aligned}
J_0(u_0) &= 2\mathbf{E}\left[\sum_{i=1}^{\infty} x_0'(i)F(i)q_0(i) + \sum_{j=0}^{\infty} u_0'(j)G(j)v(j)\right] \\
&= 2\mathbf{E}\left[\sum_{j=0}^{\infty} u_0'(j)G(j)v(j) + \sum_{i=0}^{\infty} x_0'(i+1)F(i+1)\sum_{j=0}^{i}\psi(i,j,b(\cdot,j))v(j)\right] \\
&= 2\mathbf{E}\left[\sum_{j=0}^{\infty} u_0'(j)G(j)v(j) + \sum_{j=0}^{\infty}\sum_{i=j}^{\infty} x_0'(i+1)F(i+1)\psi(i,j,b(\cdot,j))v(j)\right] \\
&= 2\mathbf{E}\sum_{j=0}^{\infty}\left[u_0'(j)G(j) + \sum_{i=j}^{\infty}\mathbf{E}_j x_0'(i+1)F(i+1)\psi(i,j,b(\cdot,j))\right]v(j).
\end{aligned}
\tag{4.23}
$$

The necessary condition $J_0(u_0) \geq 0$ for optimality of the control u_0 holds for arbitrary $v(j)$, $j \geq 0$, if and only if the expression in the square brackets in (4.23) equals zero. Thus, from (4.23) it follows that the optimal control u_0 of the control problem (4.1), (4.2) has the form (4.18), (4.19), i.e.,

$$
u_0(j) = -\sum_{i=j}^{\infty} G^{-1}(j)\psi'(i,j,b(\cdot,j))F(i+1)\mathbf{E}_j x_0(i+1), \qquad j \geq 0. \tag{4.24}
$$

The proof is completed.

Remark 4.5 If $F(N) = F$ for some $N > 0$ and $F(i) = 0$ for $i \neq N$ then from (4.23), (4.24) the representation (2.53) follows.

To study some useful properties of the control (4.18), (4.19) let us suppose that

$$
B = \sup_{i \geq 0} \sum_{j=0}^{i} |b(i,j)| < \infty, \tag{4.25}
$$

$$
\hat{B} = \sup_{j \geq 0} \sum_{i=j}^{\infty} |b(i,j)| < \infty, \tag{4.26}
$$

$$
G_1 = \sup_{j \geq 0} |G^{-1}(j)| < \infty, \tag{4.27}
$$

$$
F = \sup_{j \geq 0} |F(j)| < \infty, \tag{4.28}
$$

and put

$$
\gamma = \sup_{i \geq 0} \sum_{j=0}^{i} |\gamma(i,j)|, \qquad \hat{\gamma} = \sup_{j \geq 0} \sum_{i=j}^{\infty} |\gamma(i,j)|. \tag{4.29}
$$

Lemma 4.4 *If the conditions* (4.8), (4.9), (4.25)–(4.28) *hold then*

$$\gamma \le G_1 FB(1+R), \qquad \hat{\gamma} \le G_1 F\hat{B}\left(1+\hat{R}\right), \tag{4.30}$$

and the optimal control (4.18), (4.19) *satisfies the conditions*

$$\|u_0\| \le \hat{\gamma}\|x_0\|, \qquad S(u_0) \le \hat{\gamma}\gamma S(x_0). \tag{4.31}$$

Proof From the proof of Lemma 4.2, it follows that via the conditions (4.8), (4.9) the conditions (4.4), (4.5) hold. Via (4.19), (4.27), (4.28), (2.52), (4.25), (4.4) it follows that

$$\begin{aligned}
\sum_{j=0}^{i}|\gamma(i,j)| &\le G_1 F\left[\sum_{j=0}^{i}|b(i,j)| + \sum_{j=0}^{i}\sum_{k=j+1}^{i}|R(i,k)||b(k-1,j)|\right] \\
&\le G_1 F\left[B + \sum_{k=1}^{i}|R(i,k)|\sum_{j=0}^{k-1}|b(k-1,j)|\right] \\
&\le G_1 FB(1+R).
\end{aligned}$$

From this and (4.29) the first condition (4.30) follows.

Similarly, via (4.19), (4.27), (4.28), (2.52), (4.26), (4.5) it follows that

$$\begin{aligned}
\sum_{i=j}^{\infty}|\gamma(i,j)| &\le G_1 F\left[\sum_{i=j}^{\infty}|b(i,j)| + \sum_{i=j}^{\infty}\sum_{k=j+1}^{i}|R(i,k)||b(k-1,j)|\right] \\
&\le G_1 F\left[\hat{B} + \sum_{k=j+1}^{\infty}\sum_{i=k}^{\infty}|R(i,k)||b(k-1,j)|\right] \\
&\le G_1 F\hat{B}\left(1+\hat{R}\right).
\end{aligned}$$

From this and (4.29) the second condition (4.30) follows.

From (4.18) it follows that

$$\begin{aligned}
\mathbf{E}|u_0(j)|^2 &\le \mathbf{E}\left(\sum_{i=j}^{\infty}|\gamma(i,j)||\mathbf{E}_j x_0(i+1)|\right)^2 \\
&\le \sum_{k=j}^{\infty}|\gamma(k,j)|\sum_{i=j}^{\infty}|\gamma(i,j)|\mathbf{E}|\mathbf{E}_j x_0(i+1)|^2.
\end{aligned}$$

Via $\mathbf{E}|\mathbf{E}_j x_0(l)|^2 \le \mathbf{E}|x_0(l)|^2, j < l$, and (4.29) we have

$$\begin{aligned}\mathbf{E}|u_0(j)|^2 &\le \sum_{k=j}^{\infty}|\gamma(k,j)|\sum_{i=j}^{\infty}|\gamma(i,j)|\mathbf{E}|x_0(i+1)|^2\\ &\le \hat{\gamma}^2\|x\|^2.\end{aligned} \tag{4.32}$$

From this the first condition (4.31) follows.

Summing (4.32) over $j \ge 0$, by virtue of (4.29) we obtain

$$\begin{aligned}S(u_0) &\le \sum_{j=0}^{\infty}\sum_{k=j}^{\infty}|\gamma(k,j)|\sum_{i=j}^{\infty}|\gamma(i,j)|\mathbf{E}|x_0(i+1)|^2\\ &\le \hat{\gamma}\sum_{j=0}^{\infty}\sum_{i=j}^{\infty}|\gamma(i,j)|\mathbf{E}|x_0(i+1)|^2\\ &\le \hat{\gamma}\sum_{i=0}^{\infty}\sum_{j=0}^{i}|\gamma(i,j)|\mathbf{E}|x_0(i+1)|^2\\ &\le \hat{\gamma}\gamma S(x_0).\end{aligned}$$

So, the second condition (4.31) is obtained too. The proof is completed.

Remark 4.6 Let the conditions (4.8), (4.9), (4.25)–(4.28) and

$$G = \sup_{j\ge 0}|G(j)| < \infty \tag{4.33}$$

hold. Then from Lemma 4.4 in particular follows that if the solution $x_0(j)$ of the Eq. (4.1) by the optimal control $u_0(j)$ is mean square summable, i.e., $S(x_0) < \infty$, then the control $u_0(j)$ is mean square summable too, i.e., $S(u_0) < \infty$. By this the performance functional (4.2) is bounded since from (4.2), (4.28), (4.33), (4.31) it follows that

$$\begin{aligned}J(u_0) &\le FS(x_0) + GS(u_0)\\ &\le (F + G\hat{\gamma}\gamma)S(x_0)\\ &< \infty.\end{aligned}$$

4.1.2.2 Synthesis of the Optimal Control

To construct a synthesis of the optimal control of the control problem (4.1), (4.2), it is necessary to represent the expression $\mathbf{E}_j x_0(i)$, $j < i$, from (4.24) in the form of a functional of the $x_0(k)$ for $k \le j$. Thereto put

$$\beta(i,j) = \mathbf{E}_j\eta(i+1) - \eta(j), \qquad i \ge j, \tag{4.34}$$

$$\zeta_0(i,j+1) = \psi(i,j,\beta(\cdot,j+1)) + \psi(i,j,I)x_0(j+1) + \sum_{k=0}^{j} \psi(i,j,a_j(\cdot,k))x_0(k), \qquad i > j, \tag{4.35}$$

$$\mu_0^j(k) = \sum_{i=k}^{\infty} \gamma(i,k)\zeta_0(i,j), \qquad k \geq j, \tag{4.36}$$

$$\mu_1^j(k,m) = \sum_{i=k}^{\infty} \gamma(i,k)\psi(i,j-1,b_{j-1}(\cdot,m)), \quad k \geq j > m, \tag{4.37}$$

$$\mu_2^j(k,m) = \sum_{i=k\vee m}^{\infty} \gamma(i,k)\psi(i,m,b(\cdot,m)), \quad k,\ m \geq j, \tag{4.38}$$

where $k \vee m = \max(k,m)$, $\psi(i,j,f(\cdot))$, $\gamma(i,k)$ and $a_j(i,k)$, $b_j(i,k)$ are defined by (2.52), (4.19) and (2.61) respectively, and consider the properties of the functions $\mu_0^j(k)$, $\mu_1^j(k,m)$, $\mu_2^j(k,m)$.

Lemma 4.5 *Let the sequences $x_0(i)$ and $\eta(i)$ are uniformly mean square bounded and the conditions* (4.5), (4.8), (4.26)–(4.28) *hold. Then*

$$\mathbf{E}|\mu_0^j(k)|^2 \leq 3\hat{\gamma}^2(1+R)^2\left(4\|\eta\|^2 + 5\|x_0\|^2\right), \tag{4.39}$$

and

$$|\mu_1^j(k,m)| \leq \hat{\gamma}(1+R)\hat{B}, \qquad |\mu_2^j(k,m)| \leq \hat{\gamma}(1+R)\hat{B}. \tag{4.40}$$

Proof From (4.36) it follows that

$$\mathbf{E}|\mu_0^j(k)|^2 \leq \sum_{m=k}^{\infty} |\gamma(m,k)| \sum_{l=k}^{\infty} |\gamma(l,k)|\mathbf{E}|\zeta_0(l,j)|^2. \tag{4.41}$$

From (4.35) via (1.30) we have

$$\begin{aligned}
\mathbf{E}|\zeta_0(i,j+1)|^2 &\leq 3\Big[\mathbf{E}|\psi(i,j,\beta(\cdot,j+1))|^2 + |\psi(i,j,I)|^2\mathbf{E}|x_0(j+1)|^2 \\
&\quad + \textstyle\sum_{l=0}^{j} |\psi(i,j,a_j(\cdot,l))| \sum_{k=0}^{j} |\psi(i,j,a_j(\cdot,k))|\mathbf{E}|x_0(k)|^2\Big] \\
&\leq 3\Big[\mathbf{E}|\psi(i,j,\beta(\cdot,j+1))|^2 \\
&\quad + \Big(|\psi(i,j,I)|^2 + \big(\textstyle\sum_{k=0}^{j} |\psi(i,j,a_j(\cdot,k))|\big)^2\Big)\|x_0\|^2\Big].
\end{aligned} \tag{4.42}$$

Note that via the proof of Lemma 4.2 the condition (4.8) implies (4.4). From (2.52) and (4.4) it follows that

$$|\psi(i,j,I)| = \left| I + \sum_{k=j+1}^{i} R(i,k) \right| \le 1 + R. \tag{4.43}$$

Via Lemma 1.8 for some $\alpha > 0$ from (2.52) and (4.4) we obtain also

$$\begin{aligned}\mathbf{E}|\psi(i,j,\beta(\cdot,j+1))|^2 &\le (1+\alpha)\mathbf{E}|\beta(i,j+1)|^2 \\ &\quad + \left(1+\frac{1}{\alpha}\right) R \sum_{k=j+1}^{i} \big|R(i,k)\big|\mathbf{E}|\beta(k-1,j+1)|^2. \end{aligned} \tag{4.44}$$

From (4.34) it follows that

$$\begin{aligned}\mathbf{E}|\beta(i,j+1)|^2 &= \mathbf{E}|\mathbf{E}_{j+1}[\eta(i+1) - \eta(j+1)]|^2 \\ &\le \mathbf{E}[\mathbf{E}_{j+1}|\eta(i+1)| + |\eta(j+1)|]^2 \\ &\le 4\|\eta\|^2. \end{aligned} \tag{4.45}$$

So, from (4.44) and (4.45) for $\alpha = R$ we have

$$\mathbf{E}|\psi(i,j,\beta(\cdot,j+1))|^2 \le 4(1+R)^2\|\eta\|^2. \tag{4.46}$$

Via (2.52), (4.8) and (4.4) we have

$$\begin{aligned}\sum_{k=0}^{j} |\psi(i,j,a_j(\cdot,k))| &\le \sum_{k=0}^{j} \left[|a_j(i,k)| + \sum_{m=j+1}^{i} |R(i,m)||a_j(m-1,k)| \right] \\ &\le \sum_{k=0}^{j} \big(|a(i,k)| + |a(j,k)|\big) \\ &\quad + \sum_{m=j+1}^{i} \big|R(i,m)\big| \sum_{k=0}^{j} (|a(m-1,k)| + |a(j,k)|) \\ &\le 2(1+R). \end{aligned} \tag{4.47}$$

So, from (4.42) via (4.43), (4.46), (4.47) it follows that

$$\begin{aligned}\mathbf{E}|\zeta_0(i,j+1)|^2 &\le 3[4(1+R)^2\|\eta\|^2 + ((1+R)^2 + 4(1+R)^2)\|x_0\|^2] \\ &= 3(1+R)^2(4\|\eta\|^2 + 5\|x_0\|^2). \end{aligned}$$

As a result from this and (4.41), (4.29) we obtain (4.39). From the lemma conditions and (4.30) the boundedness of the right-hand side of (4.39) follows.

From (4.37) it follows that

$$|\mu_1^j(k,m)| \leq \sum_{i=k}^{\infty} |\gamma(i,k)| \left[|b_{j-1}(i,m)| + \sum_{l=j}^{i} |R(i,l)||b_{j-1}(l-1,m)| \right].$$

Via (2.61) and (4.26) we have

$$|b_{j-1}(i,m)| \leq |b(i,m)| + |b(j-1,m)| \leq \sum_{i=m}^{\infty} |b(i,m)| \leq \hat{B}.$$

Therefore, by virtue of (4.29) and (4.4) we obtain the first condition (4.40).
Similarly, from (4.38) we obtain the second condition (4.40):

$$\begin{aligned} |\mu_2^j(k,m)| \leq & \sum_{i=k\vee m}^{\infty} |\gamma(i,k)| \Bigg[|b(i,m)| \\ & + \sum_{l=m+1}^{i} |R(i,l)||b(l-1,m)| \Bigg] \\ \leq & \hat{\gamma}(1+R)\hat{B}. \end{aligned}$$

Note also that from the conditions (4.5), (4.26)–(4.28), (4.30) it follows that $\hat{\gamma} < \infty$. The proof is completed.

Assume now that the kernel $\mu_2^j(k,m)$ has the Fredholm resolvent $R_2^j(k,m)$, $k,m \geq j$, i.e., via (1.25)

$$R_2^j(k,m) = \mu_2^j(k,m) + \sum_{i=j}^{\infty} R_2^j(k,i)\mu_2^j(i,m), \tag{4.48}$$

and this resolvent satisfies the conditions

$$R_2 = \sup_{i \geq 0} \sum_{j=0}^{i} |R_2^j(j,i)| < \infty, \tag{4.49}$$

$$\hat{R}_2 = \sup_{j \geq 0} \sum_{i=j}^{\infty} |R_2^j(j,i)| < \infty. \tag{4.50}$$

For arbitrary function $f(k)$, $j \leq k \leq i$, put

$$\Psi(i,j,f(\cdot)) = f(j) + \sum_{k=j}^{i} R_2^j(j,k)f(k). \tag{4.51}$$

For $i = \infty$ instead of $\Psi(\infty, j, f(\cdot))$ we will write $\Psi(j, f(\cdot))$. Put also

$$\Psi = \sup_{i\geq 0} \sum_{j=0}^{i} |\Psi(i,j,\gamma(i,\cdot))|, \tag{4.52}$$

$$\hat{\Psi} = \sup_{j\geq 0} \sum_{i=j}^{\infty} |\Psi(i,j,\gamma(i,\cdot))|. \tag{4.53}$$

Lemma 4.6 *Let the conditions* (4.8), (4.9), (4.25)–(4.28), (4.49), (4.50) *hold. Then*

$$\Psi \leq \gamma(1+R_2), \qquad \hat{\Psi} \leq \hat{\gamma}(1+\hat{R}_2). \tag{4.54}$$

Proof From (4.51), (4.29), (4.49) we obtain the first condition (4.54):

$$\begin{aligned}
\sum_{j=0}^{i} |\Psi(i,j,\gamma(i,\cdot))| &\leq \sum_{j=0}^{i} |\gamma(i,j)| + \sum_{j=0}^{i}\sum_{k=j}^{i} |R_2^j(j,k)||\gamma(i,k)| \\
&\leq \gamma + \sum_{k=0}^{i}\sum_{j=0}^{k} |R_2^j(j,k)||\gamma(i,k)| \\
&\leq \gamma(1+R_2).
\end{aligned}$$

Similarly, from (4.51), (4.29), (4.50) the second condition (4.54) follows too:

$$\begin{aligned}
\sum_{i=j}^{\infty} |\Psi(i,j,\gamma(i,\cdot))| &\leq \sum_{i=j}^{\infty} |\gamma(i,j)| + \sum_{i=j}^{\infty}\sum_{k=j}^{i} |R_2^j(j,k)||\gamma(i,k)| \\
&\leq \sum_{i=j}^{\infty} |\gamma(i,j)| + \sum_{l=j}^{\infty} |R_2^j(j,k)| \sum_{i=k}^{\infty} |\gamma(i,k)| \\
&\leq \hat{\gamma}(1+\hat{R}_2).
\end{aligned}$$

The proof is completed.

Suppose now that the resolvent $R_1(j,m)$ of the kernel $\Psi(j+1, \mu_1^{j+1}(\cdot,m))$ satisfies the conditions

$$R_1 = \sup_{i\geq 0} \sum_{j=0}^{i} |R_1(i,j)| < \infty, \tag{4.55}$$

$$\hat{R}_1 = \sup_{j\geq 0} \sum_{i=j}^{\infty} |R_1(i,j)| < \infty, \tag{4.56}$$

and put for $j \geq 0$

$$\begin{aligned}
\beta_0(j) &= \sum_{m=j}^{\infty} \Psi(m,j,\gamma(m,\cdot))\psi(m,j-1,\beta(\cdot,j)),\\
\beta_1(j) &= \sum_{m=j}^{\infty} \Psi(m,j,\gamma(m,\cdot))\psi(m,j-1,I),\\
\beta_2(j) &= \sum_{m=j}^{\infty} \Psi(m,j,\gamma(m,\cdot))\psi(m,j-1,a_{j-1}(\cdot,k)),
\end{aligned} \tag{4.57}$$

$$\alpha_0(j+1) = \beta_0(j+1) + \sum_{m=0}^{j} R_1(j,m)\beta_0(m), \qquad \alpha_1(j+1) = \beta_1(j+1),$$

$$\alpha_2(j+1,k) = R_1(j,k)\beta_1(k) + \beta_2(j+1,k) + \sum_{m=k+1}^{j} R_1(j,m)\beta_2(m,k). \tag{4.58}$$

Besides put

$$\alpha_0(0) = \sum_{l=0}^{\infty} Q(l)\psi(l,0,\mathbf{E}_0\eta(\cdot+1)), \qquad \alpha_1(0) = \sum_{l=0}^{\infty} Q(l)R(l,0),$$

$$Q(l) = \gamma(l,0) + \sum_{i=0}^{\infty} \gamma(i,0)R_3(i,l), \tag{4.59}$$

where $R_3(i,l)$ is the Fredholm resolvent of the kernel

$$A(i,l) = -\sum_{j=0}^{\min\{i,l\}} \psi(i,j,b(\cdot,j))G^{-1}(j)\psi'(l,j,b(\cdot,j))F(l+1). \tag{4.60}$$

Theorem 4.1 *Let the conditions* (4.4), (4.5), (4.25)–(4.28), (4.49), (4.50), (4.55), (4.56) *hold. Then the optimal control of the control problem* (4.1), (4.2) *has the form*

$$\begin{aligned}
u_0(0) &= \alpha_0(0) + \alpha_1(0)x_0(0),\\
u_0(j+1) &= \alpha_0(j+1) + \alpha_1(j+1)x_0(j+1)\\
&\quad + \sum_{k=0}^{j} \alpha_2(j+1,k)x_0(k), \qquad j \geq 0.
\end{aligned} \tag{4.61}$$

Proof From Lemma 4.3 and (4.18) it follows that

$$\mathbf{E}_{j+1}u_0(k) = \sum_{i=k}^{\infty} \gamma(i,k)\mathbf{E}_{j+1}x_0(i+1), \qquad k > j. \tag{4.62}$$

Using (4.1) for $x = x_0$, $u = u_0$ and (4.34) for $i > j$, we obtain

$$\begin{aligned}
\mathbf{E}_{j+1}x_0(i+1) &= x_0(j+1) + \beta(i,j+1) \\
&\quad + \sum_{k=j+1}^{i} a(i,k)\mathbf{E}_{j+1}x_0(k) + \sum_{k=j+1}^{i} b(i,k)\mathbf{E}_{j+1}u_0(k) \\
&\quad + \sum_{k=0}^{j} a_j(i,k)x_0(k) + \sum_{k=0}^{j} b_j(i,k)u_0(k).
\end{aligned}$$

Let $R(i,j)$ be the resolvent of the kernel $a(i,j)$. Then from the obtained equality we get

$$\begin{aligned}
\mathbf{E}_{j+1}x_0(i+1) &= x_0(j+1) + \beta(i,j+1) + \sum_{k=0}^{j} a_j(i,k)x_0(k) \\
&\quad + \sum_{k=0}^{j} b_j(i,k)u_0(k) + \sum_{k=j+1}^{i} b(i,k)\mathbf{E}_{j+1}u_0(k) \\
&\quad + \sum_{k=j+1}^{i} R(i,k)\Bigg[x_0(j+1) + \beta(k-1,j+1) \\
&\quad + \sum_{l=0}^{j} a_j(k-1,l)x_0(l) + \sum_{l=0}^{j} b_j(k-1,l)u_0(l) \\
&\quad + \sum_{l=j+1}^{k-1} b(k-1,l)\mathbf{E}_{j+1}u_0(l)\Bigg].
\end{aligned} \tag{4.63}$$

Using the functional $\psi(i,j,f(\cdot))$ and (4.35), rewrite (4.63) in the form

$$\begin{aligned}
\mathbf{E}_{j+1}x_0(i+1) &= \zeta_0(i,j+1) + \sum_{k=0}^{j} \psi(i,j,b_j(\cdot,k))u_0(k) \\
&\quad + \sum_{k=j+1}^{i} \psi(i,k,b(\cdot,k))\mathbf{E}_{j+1}u_0(k).
\end{aligned} \tag{4.64}$$

Substituting (4.64) into (4.62) and using (4.36), (4.37), we obtain

$$
\begin{aligned}
\mathbf{E}_{j+1}u_0(k) &= \sum_{i=k}^{\infty}\gamma(i,k)\Big[\zeta_0(i,j+1)+\sum_{m=0}^{j}\psi(i,j,b_j(\cdot,m))u_0(m)\\
&\quad+\sum_{m=j+1}^{i}\psi(i,m,b(\cdot,m))\mathbf{E}_{j+1}u_0(m)\Big]\\
&= \mu_0^{j+1}(k)+\sum_{m=0}^{j}\mu_1^{j+1}(k,m)u_0(m)\\
&\quad+\sum_{i=k}^{\infty}\gamma(i,k)\sum_{m=j+1}^{i}\psi(i,m,b(\cdot,m))\mathbf{E}_{j+1}u_0(m). \qquad (4.65)
\end{aligned}
$$

Note that for arbitrary kernel $p(i,m)$ we have

$$
\begin{aligned}
\sum_{i=k}^{\infty}\sum_{m=j+1}^{i}p(i,m) &= \sum_{i=k}^{\infty}\Big[\sum_{m=j+1}^{k-1}p(i,m)+\sum_{m=k}^{i}p(i,m)\Big]\\
&= \sum_{m=j+1}^{k-1}\sum_{i=k}^{\infty}p(i,m)+\sum_{m=k}^{\infty}\sum_{i=m}^{\infty}p(i,m)\\
&= \sum_{m=j+1}^{\infty}\sum_{i=k\vee m}^{\infty}p(i,m),
\end{aligned}
$$

where $k\vee m=\max(k,m)$. Therefore, from (4.65) via (4.38) we obtain

$$
\begin{aligned}
\mathbf{E}_{j+1}u_0(k) &= \mu_0^{j+1}(k)+\sum_{m=0}^{j}\mu_1^{j+1}(k,m)u_0(m)\\
&\quad+\sum_{m=j+1}^{\infty}\mu_2^{j+1}(k,m)\mathbf{E}_{j+1}u_0(m). \qquad (4.66)
\end{aligned}
$$

Let $R_2^{j+1}(k,l)$ be the Fredholm resolvent of the kernel $\mu_2^{j+1}(k,l)$. So, via (1.23) from (4.66) it follows that

$$
\begin{aligned}
\mathbf{E}_{j+1}u_0(k) &= \mu_0^{j+1}(k)+\sum_{m=0}^{j}\mu_1^{j+1}(k,m)u_0(m)\\
&\quad+\sum_{l=j+1}^{\infty}R_2^{j+1}(k,l)\Big[\mu_0^{j+1}(l)+\sum_{m=0}^{j}\mu_1^{j+1}(l,m)u_0(m)\Big]
\end{aligned}
$$

$$
\begin{aligned}
&= \mu_0^{j+1}(k) + \sum_{l=j+1}^{\infty} R_2^{j+1}(k,l)\mu_0^{j+1}(l) \\
&\quad + \sum_{m=0}^{j}\left[\mu_1^{j+1}(k,m) + \sum_{l=j+1}^{\infty} R_2^{j+1}(k,l)\mu_1^{j+1}(l,m)\right]u_0(m).
\end{aligned}
$$

Putting here $k = j+1$ via (4.51), we obtain

$$
u_0(j+1) = \Psi(j+1, \mu_0^{j+1}(\cdot)) + \sum_{m=0}^{j} \Psi(j+1, \mu_1^{j+1}(\cdot, m))u_0(m).
$$

Since $R_1(j,m)$ is the resolvent of the kernel $\Psi(j+1, \mu_1^{j+1}(\cdot, m))$ then from this it follows that

$$
u_0(j+1) = \Psi(j+1, \mu_0^{j+1}(\cdot)) + \sum_{m=0}^{j} R_1(j,m)\Psi(m, \mu_0^{m}(\cdot)). \tag{4.67}
$$

Using (4.51), (4.36), we have

$$
\begin{aligned}
\Psi(j+1, \mu_0^{j+1}(\cdot)) &= \mu_0^{j+1}(j+1) + \sum_{l=j+1}^{\infty} R_2^{j+1}(j+1,l)\mu_0^{j+1}(l) \\
&= \sum_{m=j+1}^{\infty} \gamma(m, j+1)\zeta_0(m, j+1) \\
&\quad + \sum_{l=j+1}^{\infty} R_2^{j+1}(j+1,l)\sum_{m=l}^{\infty}\gamma(m,l)\zeta_0(m,j+1) \\
&= \sum_{m=j+1}^{\infty} \gamma(m, j+1)\zeta_0(m, j+1) \\
&\quad + \sum_{m=j+1}^{\infty}\sum_{l=j+1}^{m} R_2^{j+1}(j+1,l)\gamma(m,l)\zeta_0(m,j+1) \\
&= \sum_{m=j+1}^{\infty}\left[\gamma(m,j+1)\right. \\
&\quad \left. + \sum_{l=j+1}^{m} R_2^{j+1}(j+1,l)\gamma(m,l)\right]\zeta_0(m,j+1).
\end{aligned}
$$

From this via (4.51) it follows that

$$\Psi(j+1,\mu_0^{j+1}(\cdot)) = \sum_{m=j+1}^{\infty} \Psi(m,j+1,\gamma(m,\cdot))\zeta_0(m,j+1).$$

Substituting (4.35) into the obtained equality and using (4.57), we obtain

$$\begin{aligned}\Psi(j+1,\mu_0^{j+1}(\cdot)) &= \sum_{m=j+1}^{\infty} \Psi(m,j+1,\gamma(m,\cdot))\Big[\psi(m,j,\beta(\cdot,j+1)) \\ &\quad + \psi(m,j,I)x_0(j+1) \\ &\quad + \sum_{k=0}^{j}\psi(m,j,a_j(\cdot,k))x_0(k)\Big] \\ &= \beta_0(j+1) + \beta_1(j+1)x_0(j+1) + \sum_{k=0}^{j}\beta_2(j+1,k)x_0(k).\end{aligned} \tag{4.68}$$

Substituting (4.68) into (4.67), via (4.58) we have

$$\begin{aligned}u_0(j+1) &= \beta_0(j+1) + \beta_1(j+1)x_0(j+1) + \sum_{k=0}^{j}\beta_2(j+1,k)x_0(k) \\ &\quad + \sum_{m=0}^{j} R_1(j,m)\left[\beta_0(m) + \beta_1(m)x_0(m) + \sum_{k=0}^{m-1}\beta_2(m,k)x_0(k)\right] \\ &= \beta_0(j+1) + \sum_{m=0}^{j} R_1(j,m)\beta_0(m) \\ &\quad + \beta_1(j+1)x_0(j+1) + \sum_{m=0}^{j} R_1(j,m)\beta_1(m)x_0(m) \\ &\quad + \sum_{k=0}^{j}\beta_2(j+1,k)x_0(k) + \sum_{m=0}^{j} R_1(j,m)\sum_{k=0}^{m-1}\beta_2(m,k)x_0(k) \\ &= \beta_0(j+1) + \sum_{m=0}^{j} R_1(j,m)\beta_0(m) + \beta_1(j+1)x_0(j+1)\end{aligned}$$

$$+\sum_{k=0}^{j}\left[R_1(j,k)\beta_1(k)+\beta_2(j+1,k)+\sum_{m=k+1}^{j}R_1(j,m)\beta_2(m,k)\right]x_0(k)$$

$$=\alpha_0(j+1)+\alpha_1(j+1)x_0(j+1)+\sum_{k=0}^{j}\alpha_2(j+1,k)x_0(k).$$

So, the second equality (4.61) is obtained.

To calculate $u_0(0)$ note that via (4.18)

$$u_0(0)=\sum_{i=0}^{\infty}\gamma(i,0)\mathbf{E}_0x_0(i+1). \tag{4.69}$$

Using the resolvent $R(i,j)$ of the kernel $a(i,j)$, from (4.1) for $x=x_0$ and $u=u_0$ we have

$$x_0(i+1)=\eta(i+1)+\sum_{j=0}^{i}b(i,j)u_0(j)$$
$$+\sum_{k=0}^{i}R(i,k)\left(\eta(k)+\sum_{j=0}^{k-1}b(k-1,j)u_0(j)\right)$$
$$=\eta(i+1)+\sum_{k=1}^{i}R(i,k)\eta(k)+R(i,0)x_0(0)$$
$$+\sum_{j=0}^{i}\left(b(i,j)+\sum_{k=j+1}^{i}R(i,k)b(k-1,j)\right)u_0(j).$$

From this via (2.52) it follows that

$$\mathbf{E}_0x_0(i+1)=y(i)+\sum_{j=0}^{i}\psi(i,j,b(\cdot,j))\mathbf{E}_0u_0(j), \tag{4.70}$$

where

$$y(i)=\psi(i,0,\mathbf{E}_0\eta(\cdot+1))+R(i,0)x_0(0),\qquad i\geq 0. \tag{4.71}$$

Via (4.18) the sum in (4.70) can be transformed by the following way

$$\sum_{j=0}^{i}\psi(i,j,b(\cdot,j))\mathbf{E}_0u_0(j)=\sum_{j=0}^{i}\psi(i,j,b(\cdot,j))\sum_{l=j}^{\infty}\gamma(l,j)\mathbf{E}_0x_0(l+1)$$

$$
\begin{aligned}
&= \sum_{j=0}^{i} \psi(i,j,b(\cdot,j)) \sum_{l=j}^{i} \gamma(l,j)\mathbf{E}_0 x_0(l+1) \\
&\quad + \sum_{j=0}^{i} \psi(i,j,b(\cdot,j)) \sum_{l=i+1}^{\infty} \gamma(l,j)\mathbf{E}_0 x_0(l+1) \\
&= \sum_{l=0}^{i} \sum_{j=0}^{l} \psi(i,j,b(\cdot,j))\gamma(l,j)\mathbf{E}_0 x_0(l+1) \\
&\quad + \sum_{l=i+1}^{\infty} \sum_{j=0}^{i} \psi(i,j,b(\cdot,j))\gamma(l,j)\mathbf{E}_0 x_0(l+1) \\
&= \sum_{l=0}^{\infty} \sum_{j=0}^{\min\{i,l\}} \psi(i,j,b(\cdot,j))\gamma(l,j)\mathbf{E}_0 x_0(l+1).
\end{aligned}
$$

As a result from this, (4.70) and (4.19) we obtain

$$
\mathbf{E}_0 x_0(i+1) = y(i) + \sum_{l=0}^{\infty} A(i,l)\mathbf{E}_0 x_0(l+1),
$$

where the kernel $A(i,l)$ is defined by (4.60).

Using the Fredholm resolvent $R_3(i,l)$ of the kernel $A(i,l)$, from this via (1.23) we have

$$
\mathbf{E}_0 x_0(i+1) = y(i) + \sum_{l=0}^{\infty} R_3(i,l)y(l).
$$

Substituting this into (4.69), via (4.59), (4.71) we obtain

$$
\begin{aligned}
u_0(0) &= \sum_{i=0}^{\infty} \gamma(i,0)\left(y(i) + \sum_{l=0}^{\infty} R_3(i,l)y(l) \right) \\
&= \sum_{l=0}^{\infty} \gamma(l,0)y(l) + \sum_{i=0}^{\infty} \gamma(i,0) \sum_{l=0}^{\infty} R_3(i,l)y(l) \\
&= \sum_{l=0}^{\infty} Q(l)y(l) = \alpha_0(0) + \alpha_1(0)x_0(0).
\end{aligned}
$$

The proof is completed.

4.1.3 Stability by the Optimal Control

Let us obtain sufficient conditions for mean square stability and mean square summability of the solution $x_0(j)$ of the Eq. (4.1) by the optimal control $u_0(j)$ given by (4.61). Consider now the Eq. (4.1) by the optimal control (4.61). For this goal substitute (4.61) into (4.1):

$$
\begin{aligned}
x_0(i+1) &= \eta(i+1) + \sum_{j=0}^{i} a(i,j)x_0(j) \\
&\quad + \sum_{j=0}^{i} b(i,j)\left[\alpha_0(j) + \alpha_1(j)x_0(j) + \sum_{k=0}^{j-1} \alpha_2(j,k)x_0(k)\right] \\
&= \eta(i+1) + \sum_{j=0}^{i} b(i,j)\alpha_0(j) \\
&\quad + \sum_{j=0}^{i} [a(i,j) + b(i,j)\alpha_1(j)]x_0(j) + \sum_{j=0}^{i} b(i,j)\sum_{k=0}^{j-1} \alpha_2(j,k)x_0(k).
\end{aligned}
$$

Changing the order of summation in the last summand, we obtain

$$
x_0(i+1) = \eta_1(i+1) + \sum_{j=0}^{i} a_1(i,j)x_0(j), \tag{4.72}
$$

where

$$
\eta_1(i+1) = \eta(i+1) + \sum_{j=0}^{i} b(i,j)\alpha_0(j), \quad i \in Z, \quad \eta_1(0) = \eta(0), \tag{4.73}
$$

$$
a_1(i,j) = a(i,j) + b(i,j)\alpha_1(j) + \sum_{k=j+1}^{i} b(i,k)\alpha_2(k,j). \tag{4.74}
$$

Here $\alpha_0(j)$, $\alpha_1(j)$, and $\alpha_2(j)$ are defined in (4.58), (4.59).

Lemma 4.7 *Let the conditions* (4.4), (4.5), (4.25)–(4.28), (4.50), (4.55) *hold. Then if the sequence* $\eta(i)$ *is uniformly mean square bounded then the sequence* $\eta_1(i)$ *is uniformly mean square bounded too.*

Proof From (4.73) via (1.31) and (4.25) it follows that for arbitrary $\alpha_1 > 0$

$$\mathbf{E}|\eta_1(i+1)|^2 \leq (1+\alpha_1)\mathbf{E}|\eta(i+1)|^2 + \left(1+\frac{1}{\alpha_1}\right) B \sum_{j=0}^{i} |b(i,j)|\mathbf{E}|\alpha_0(j)|^2. \quad (4.75)$$

Similarly, from (4.58) via (4.55) for arbitrary $\alpha_2 > 0$ we have

$$\mathbf{E}|\alpha_0(j+1)|^2 \leq (1+\alpha_2)\mathbf{E}|\beta_0(j+1)|^2 + \left(1+\frac{1}{\alpha_2}\right) R_1 \sum_{m=0}^{j} |R_1(j,m)|\mathbf{E}|\beta_0(m)|^2. \quad (4.76)$$

From (4.57) via (4.46), (4.53), (4.54) we obtain

$$\begin{aligned}\mathbf{E}|\beta_0(j+1)|^2 &\leq \hat{\Psi} \sum_{m=j+1}^{\infty} |\Psi(m,j+1,\gamma(m,\cdot))|\mathbf{E}|\psi(m,j,\beta(\cdot,j+1))|^2 \\ &\leq 4\hat{\gamma}^2(1+R)^2(1+\hat{R}_2)^2\|\eta\|^2.\end{aligned}$$

As a result from this and (4.76) by $\alpha_2 = R_1$ and $D = 2\hat{\gamma}(1+R)(1+R_1)(1+\hat{R}_2)$ we get

$$\mathbf{E}|\alpha_0(j+1)|^2 \leq D^2\|\eta\|^2.$$

So, from (4.75) by $\alpha_1 = BD$ it follows that

$$\|\eta_1(i+1)\|^2 \leq (1+BD)^2\|\eta\|^2. \quad (4.77)$$

The lemma is proven.

Lemma 4.8 *Let the conditions* (4.4), (4.5), (4.25), (4.26), (4.55), (4.56) *hold. Then if the sequence* $\eta(i)$ *is mean square summable then the sequence* $\eta_1(i)$ *is mean square summable too.*

Proof Summing the inequality (4.75) over $i \geq 0$ together with the trivial inequality $\mathbf{E}|\eta_1(0)|^2 \leq (1+\alpha_1)\mathbf{E}|\eta(0)|^2$, we obtain

$$S(\eta_1) \leq (1+\alpha_1)S(\eta) + \left(1+\frac{1}{\alpha_1}\right) B \sum_{i=0}^{\infty}\sum_{j=0}^{i} |b(i,j)|\mathbf{E}|\alpha_0(j)|^2. \quad (4.78)$$

Changing the order of summation, via (4.26) we have

$$S(\eta_1) \leq (1+\alpha_1)S(\eta) + \left(1+\frac{1}{\alpha_1}\right) B\hat{B}S(\alpha_0).$$

Therefore, for mean square summability of the sequence $\eta_1(i)$ it is enough to prove mean square summability of the sequence $\alpha_0(i)$.

From (4.76) similarly to (4.78) it follows that

$$S(\alpha_0) \le (1+\alpha_2)S(\beta_0) + \left(1+\frac{1}{\alpha_2}\right) R_1 \sum_{j=0}^{\infty}\sum_{m=0}^{j} |R_1(j,m)|\mathbf{E}|\beta_0(m)|^2.$$

Changing the order of summation and putting $\alpha_2 = \sqrt{R_1\hat{R}_1}$, via (4.56) we obtain

$$S(\alpha_0) \le \left(1+\sqrt{R_1\hat{R}_1}\right)^2 S(\beta_0). \tag{4.79}$$

So, for mean square summability of the sequence $\eta_1(i)$ it is enough to prove mean square summability of the sequence $\beta_0(i)$.

Via (4.57), (4.53) we have

$$S(\beta_0) \le \hat{\Psi} \sum_{j=0}^{\infty}\sum_{m=j}^{\infty} |\Psi(m,j,\gamma(m,\cdot))|\mathbf{E}|\psi(m,j-1,\beta(\cdot,j))|^2. \tag{4.80}$$

Via (2.52), (4.34), (4.43) and Lemma 1.8 for $\alpha_3 > 0$ we obtain

$$\begin{aligned}
\mathbf{E}|\psi(m,j-1,\beta(\cdot,j))|^2 &\le \mathbf{E}\Big[|\psi(m,j-1,\mathbf{E}_j\eta(\cdot+1))| \\
&\quad + |\psi(m,j-1,I)||\eta(j)|\Big]^2 \\
&\le (1+\alpha_3)\mathbf{E}|\psi(m,j-1,\mathbf{E}_j\eta(\cdot+1))|^2 \\
&\quad + \left(1+\frac{1}{\alpha_3}\right)(1+R)^2\mathbf{E}|\eta(j)|^2.
\end{aligned} \tag{4.81}$$

Similarly, using $\alpha_4 > 0$ and (4.4), we have

$$\begin{aligned}
\mathbf{E}|\psi(m,j-1,\mathbf{E}_j\eta(\cdot+1))|^2 &= \mathbf{E}\left|\mathbf{E}_j\eta(m+1) + \sum_{k=j}^{m} R(m,k)\mathbf{E}_j\eta(k)\right|^2 \\
&\le (1+\alpha_4)\mathbf{E}|\eta(m+1)|^2 \\
&\quad + \left(1+\frac{1}{\alpha_4}\right) R \sum_{k=j}^{m} |R(m,k)|\mathbf{E}|\eta(k)|^2.
\end{aligned} \tag{4.82}$$

So, from (4.80)–(4.82) it follows that

$$\begin{aligned}
S(\beta_0) \le \hat{\Psi} \sum_{j=0}^{\infty}\sum_{m=j}^{\infty} |\Psi(m,j,\gamma(m,\cdot))| \Big[(1+\alpha_3)\mathbf{E}|\psi(m,j-1,\mathbf{E}_j\eta(\cdot+1))|^2 \\
+ \left(1+\frac{1}{\alpha_3}\right)(1+R)^2\mathbf{E}|\eta(j)|^2\Big]
\end{aligned}$$

$$
\begin{aligned}
&\leq \hat{\Psi} \sum_{j=0}^{\infty} \sum_{m=j}^{\infty} |\Psi(m, j, \gamma(m, \cdot))| \\
&\quad \times \left[(1+\alpha_3) \left((1+\alpha_4)\mathbf{E}|\eta(m+1)|^2 + \left(1 + \frac{1}{\alpha_4}\right) \right. \right. \\
&\quad \left. \left. \times R \sum_{k=j}^{m} |R(m, k)| \mathbf{E}|\eta(k)|^2 \right) + \left(1 + \frac{1}{\alpha_3}\right) (1+R)^2 \mathbf{E}|\eta(j)|^2 \right] \\
&= \hat{\Psi} \left[(1+\alpha_3) \left[(1+\alpha_4) I_1 + \left(1 + \frac{1}{\alpha_4}\right) R I_2 \right] + \left(1 + \frac{1}{\alpha_3}\right) (1+R)^2 I_3 \right],
\end{aligned}
\tag{4.83}
$$

where

$$
\begin{aligned}
I_1 &= \sum_{j=0}^{\infty} \sum_{m=j}^{\infty} |\Psi(m, j, \gamma(m, \cdot))| \mathbf{E}|\eta(m+1)|^2, \\
I_2 &= \sum_{j=0}^{\infty} \sum_{m=j}^{\infty} |\Psi(m, j, \gamma(m, \cdot))| \sum_{k=j}^{m-1} |R(m-1, k)| \mathbf{E}|\eta(k)|^2, \\
I_3 &= \sum_{j=0}^{\infty} \sum_{m=j}^{\infty} |\Psi(m, j, \gamma(m, \cdot))| \mathbf{E}|\eta(j)|^2.
\end{aligned}
$$

Changing the order of summation, by virtue of (4.52) we obtain

$$
\begin{aligned}
I_1 &= \sum_{m=0}^{\infty} \sum_{j=0}^{m} |\Psi(m, j, \gamma(m, \cdot))| \mathbf{E}|\eta(m+1)|^2 \\
&\leq \Psi S(\eta).
\end{aligned}
$$

Similarly, using (4.52) and (4.5), we have

$$
\begin{aligned}
I_2 &= \sum_{k=0}^{\infty} \sum_{m=k+1}^{\infty} \sum_{j=0}^{m} |\Psi(m, j, \gamma(m, \cdot))| |R(m-1, k)| \mathbf{E}|\eta(k)|^2 \\
&\leq \Psi \hat{R} S(\eta)
\end{aligned}
$$

and via (4.53) $I_3 \leq \hat{\Psi} S(\eta)$.

Substituting into (4.83) the obtained estimations for I_k, $k = 1, 2, 3$ and

$$
\alpha_3 = \frac{(1+R)\sqrt{\hat{\Psi}}}{(1+\sqrt{R\hat{R}})\sqrt{\Psi}}, \qquad \alpha_4 = \sqrt{R\hat{R}},
$$

we obtain

$$S(\beta_0) \leq \hat{\Psi}\left(\left(1+\sqrt{R\hat{R}}\right)\sqrt{\Psi}+(1+R)\sqrt{\hat{\Psi}}\right)^2 S(\eta). \tag{4.84}$$

So, mean square summability of the sequence $\eta(i)$ implies mean square summability of the sequence $\beta_0(i)$ that in turn implies mean square summability of the sequence $\eta_1(i)$. The lemma is proven.

Lemma 4.9 *Let $a_1(i,j)$ be the kernel defined by (4.74) and the conditions (4.8), (4.9), (4.25), (4.26), (4.27), (4.28), (4.49), (4.50), (4.55), (4.56) hold. Then the inequality*

$$A + 3B\hat{B}G_1F(1+R)\left(1+\hat{R}\right)(1+R_1)\left(1+\hat{R}_2\right) < 1 \tag{4.85}$$

implies the condition

$$A_1 = \sup_{i\geq 0}\sum_{j=0}^{i}\left|a_1(i,j)\right| < 1, \tag{4.86}$$

and the inequality

$$\hat{A} + \hat{B}G_1F(1+R)\left(1+\hat{R}\right)\left(1+\hat{R}_1\right)\left[\hat{B}\left(1+\hat{A}\right)\left(1+\hat{R}_2\right)+\hat{A}B(1+R_2)\right] < 1 \tag{4.87}$$

implies the condition

$$\hat{A}_1 = \sup_{j\geq 0}\sum_{i=j}^{\infty}|a_1(i,j)| < 1. \tag{4.88}$$

Proof From (4.74) we have

$$\sum_{j=0}^{i}|a_1(i,j)| \leq \sum_{j=0}^{i}|a(i,j)| + \sum_{j=0}^{i}|b(i,j)||\alpha_1(j)| + \sum_{j=0}^{i}\sum_{k=j+1}^{i}|b(i,k)||\alpha_2(k,j)|.$$

From (4.58), (4.57), (4.53), (4.43), it is easy to get

$$|\alpha_1(j)| = |\beta_1(j)| \leq \hat{\Psi}(1+R). \tag{4.89}$$

Therefore, via (4.8), (4.25) we have

$$\sum_{j=0}^{i}|a_1(i,j)| \leq A + B\hat{\Psi}(1+R) + B\sup_{k\geq 1}\sum_{j=0}^{k-1}|\alpha_2(k,j)|. \tag{4.90}$$

From (4.58) using (4.55), (4.89) and changing the order of summation, we obtain

$$\begin{aligned}\sum_{k=0}^{j}|\alpha_2(j+1,k)| &\le \sum_{k=0}^{j}|R_1(j,k)||\beta_1(k)| + \sum_{k=0}^{j}|\beta_2(j+1,k)| \\ &\quad + \sum_{k=0}^{j}\sum_{m=k+1}^{j}|R_1(j,m)||\beta_2(m,k)| \\ &\le \sum_{k=0}^{j}|R_1(j,k)||\beta_1(k)| + \sum_{k=0}^{j}|\beta_2(j+1,k)| \\ &\quad + \sum_{m=1}^{j}|R_1(j,m)|\sum_{k=0}^{m-1}|\beta_2(m,k)| \\ &\le \hat{\Psi}(1+R)R_1 + (1+R_1)\sup_{j\in Z}\sum_{k=0}^{j}|\beta_2(j+1,k)|.\end{aligned}$$

From (4.57), using (4.47), (4.53), we have

$$\begin{aligned}\sum_{k=0}^{j}|\beta_2(j+1,k)| &\le \sum_{m=j+1}^{\infty}|\Psi(m,j+1,\gamma(m,\cdot))|\sum_{k=0}^{j}|\psi(m,j,a_j(\cdot,k))| \\ &\le 2\hat{\Psi}(1+R).\end{aligned}$$

Therefore,

$$\begin{aligned}\sum_{k=0}^{j}|\alpha_2(j+1,k)| &\le \hat{\Psi}(1+R)R_1 + 2\hat{\Psi}(1+R)(1+R_1) \\ &= \hat{\Psi}(1+R)(2+3R_1).\end{aligned} \tag{4.91}$$

So, via (4.90) and (4.54), (4.30) we obtain

$$\begin{aligned}\sum_{j=0}^{i}|a_1(i,j)| &\le A + B\hat{\Psi}(1+R) + B\hat{\Psi}(1+R)(2+3R_1) \\ &= A + 3B\hat{\Psi}(1+R)(1+R_1) \\ &\le A + 3B\hat{B}G_1F(1+R)(1+\hat{R})(1+R_1)(1+\hat{R}_2).\end{aligned}$$

From this and (4.85) the condition (4.86) follows.

Let us prove (4.88). From (4.74) using (4.89), (4.9), (4.26) and changing the order of summation, we obtain

$$\sum_{i=j}^{\infty}|a_1(i,j)| \leq \sum_{i=j}^{\infty}|a(i,j)| + \sum_{i=j}^{\infty}|b(i,j)||\alpha_1(j)| + \sum_{i=j}^{\infty}\sum_{k=j+1}^{i}|b(i,k)||\alpha_2(k,j)|$$
$$\leq \hat{A} + \hat{B}\hat{\Psi}(1+R) + \sum_{k=j+1}^{\infty}\sum_{i=k}^{\infty}|b(i,k)||\alpha_2(k,j)$$
$$\leq \hat{A} + \hat{B}\hat{\Psi}(1+R) + \hat{B}\sum_{k=j+1}^{\infty}|\alpha_2(k,j)|. \tag{4.92}$$

From (4.58) by virtue of (4.89), (4.56) it follows that

$$\sum_{j=k}^{\infty}|\alpha_2(j+1,k)| \leq \sum_{j=k}^{\infty}|R_1(j,k)||\beta_1(k)| + \sum_{j=k}^{\infty}|\beta_2(j+1,k)|$$
$$+ \sum_{j=k+1}^{\infty}\sum_{m=k+1}^{j}|R_1(j,m)||\beta_2(m,k)|$$
$$\leq \hat{\Psi}(1+R)\hat{R}_1 + \sum_{j=k}^{\infty}|\beta_2(j+1,k)|$$
$$+ \sum_{m=k+1}^{\infty}\sum_{j=m}^{\infty}|R_1(j,m)||\beta_2(m,k)|$$
$$\leq \hat{\Psi}(1+R)\hat{R}_1 + \left(1+\hat{R}_1\right)\sum_{j=k}^{\infty}|\beta_2(j+1,k)|. \tag{4.93}$$

From (4.57) by virtue of (2.52), (2.61) we have

$$\sum_{j=k}^{\infty}|\beta_2(j+1,k)| \leq \sum_{j=k}^{\infty}\sum_{m=j+1}^{\infty}|\Psi(m,j+1,\gamma(m,\cdot))||\psi(m,j,a_j(\cdot,k))|$$
$$\leq \sum_{j=k}^{\infty}\sum_{m=j+1}^{\infty}|\Psi(m,j+1,j+1,\gamma(m,\cdot))|$$
$$\times\Bigg(|a(m-1,k)| + |a(j,k)|$$
$$+ \sum_{l=j+1}^{m-1}|R(m-1,l)|(|a(l-1,k)| + |a(j,k)|)\Bigg)$$
$$= I_1 + I_2 + I_3 + I_4, \tag{4.94}$$

where

$$I_1 = \sum_{j=k}^{\infty} \sum_{m=j+1}^{\infty} |\Psi(m, j+1, \gamma(m, \cdot))||a(m-1, k)|,$$

$$I_2 = \sum_{j=k}^{\infty} \sum_{m=j+1}^{\infty} |\Psi(m, j+1, \gamma(m, \cdot))||a(j, k)|$$

$$I_3 = \sum_{j=k}^{\infty} \sum_{m=j+1}^{\infty} |\Psi(m, j+1, \gamma(m, \cdot))| \sum_{l=j+1}^{m-1} |R(m-1, l)||a(l-1, k)|$$

$$I_4 = \sum_{j=k}^{\infty} \sum_{m=j+1}^{\infty} |\Psi(m, j+1, \gamma(m, \cdot))| \sum_{l=j+1}^{m-1} |R(m-1, l)||a(j, k)|.$$

Changing the order of summation in I_1 and using (4.52), (4.9) and the condition (4.5) that follows from (4.9) (Lemma 4.2), we obtain

$$\begin{aligned} I_1 &= \sum_{m=k+1}^{\infty} \sum_{j=k}^{m-1} |\Psi(m, j+1, \gamma(m, \cdot))||a(m-1, k)| \\ &\leq \Psi \hat{A}, \end{aligned} \tag{4.95}$$

and similarly via (4.53), (4.9)

$$I_2 \leq \hat{\Psi} \hat{A}. \tag{4.96}$$

Changing the order of summation in I_3 and using (4.52), (4.5), (4.9) we obtain

$$I_3 \leq \sum_{l=k+1}^{\infty} \sum_{m=l+1}^{\infty} \sum_{j=k}^{l-1} |\Psi(m, j+1, j+1, \gamma(m, \cdot))||R(m-1, l)||a(l-1, k)| \quad \leq \Psi \hat{R} \hat{A}, \tag{4.97}$$

and similarly

$$I_4 \leq \hat{\Psi} R \hat{A}. \tag{4.98}$$

So, from (4.94)–(4.98) we have

$$\sum_{j=k}^{\infty} |\beta_2(j+1, k)| \leq \hat{A} \left(\hat{\Psi}(1+R) + \Psi \left(1 + \hat{R}\right)\right) \tag{4.99}$$

and via (4.93)

$$\sum_{j=k}^{\infty} |\alpha_2(j+1, k)| \leq \hat{\Psi}(1+R)\hat{R}_1 + \left(1 + \hat{R}_1\right) \hat{A} \left(\hat{\Psi}(1+R) + \Psi \left(1 + \hat{R}\right)\right). \tag{4.100}$$

As a result from (4.92) and (4.100) by virtue of (4.54) and (4.30) we obtain

$$
\begin{aligned}
\sum_{i=j}^{\infty}|a_1(i,j)| &\le \hat{A}+\hat{B}\hat{\Psi}(1+R)+\hat{B}\Big[\hat{\Psi}(1+R)\hat{R}_1+\left(1+\hat{R}_1\right)\hat{A} \\
&\quad\times\left(\hat{\Psi}(1+R)+\Psi\left(1+\hat{R}\right)\right)\Big] \\
&=\hat{A}+\hat{B}\hat{\Psi}(1+\hat{A})(1+R)(1+\hat{R}_1) \\
&\quad+\hat{B}\Psi\hat{A}(1+\hat{R})(1+\hat{R}_1) \\
&\le \hat{A}+\hat{B}G_1F(1+R)(1+\hat{R})(1+\hat{R}_1) \\
&\quad\times\left[\hat{B}(1+\hat{A})(1+\hat{R}_2)+\hat{A}B(1+R_2)\right].
\end{aligned}
$$

From this and (4.87) the condition (4.88) follows. The lemma is proven.

Theorem 4.2 *Let the conditions* (4.8), (4.9), (4.25)–(4.28), (4.49), (4.51), (4.55), (4.56), (4.85), (4.87) *hold. Then by the control* (4.61) *the solution of the equation* (4.1) *is mean square stable, is mean square summable and the performance functional* (4.2) *reaches its minimum.*

Proof The optimality of the control (4.61) of the problem (4.1) and (4.2) follows from Theorem 4.1. Since the Eq. (4.1) by the control (4.61), i.e., the Eq. (4.72), has the form (4.3), then mean square stability and mean square summability follows from Lemmas 4.1, 4.7–4.9.

Remark 4.7 Note that the conditions (4.85), (4.87) have enough difficult dependence on the kernel $b(i,j)$ since the kernels $\mu_1^j(k,m)$, $\mu_2^j(k,m)$ defined in (4.37), (4.38) and therefore $R_1, \hat{R}_1, R_2, \hat{R}_2$ defined in (4.55), (4.56), (4.49), (4.50), depend on the kernel $b(i,j)$. Nevertheless from Theorem 4.2 simple enough conclusion follows: if the kernel $a(i,j)$ in the Eq. (4.1) satisfies the conditions (4.8), (4.9) and the kernel $b(i,j)$ is small enough then the solution of the Eq. (4.1) by the control (4.61) is mean square stable and mean square summable.

4.2 Quasioptimal Stabilization Problem

Consider the stabilization problem for the quasilinear stochastic difference Volterra equation

$$
\begin{aligned}
x(i+1)&=\eta(i+1)+\sum_{j=0}^{i}a(i,j)x(j)+\sum_{j=0}^{i}b(i,j)u(j)+\varepsilon\Phi(i+1,x_{i+1}), \\
i&\ge 0,\qquad x_0=\varphi_0,\qquad \eta(0)=\varphi_0(0)-\varepsilon\Phi(0,\varphi_0),
\end{aligned}
\tag{4.101}
$$

with the quadratic performance functional (4.2). Here $\varepsilon \geq 0$, $\eta \in H$, x_i is a trajectory of the process $x(j)$ until $j \leq i$, $a(i,j)$ and $b(i,j)$ are nonrandom matrices of the dimensions $n \times n$ and $n \times m$ respectively. The functional $\Phi(i,\varphi) \in \mathbf{R}^n$ depends on the function $\varphi(j)$ for $j = 0, \ldots, i$, $\varphi \in \tilde{H}$, and satisfies the condition

$$|\Phi(i,\varphi)| \leq \sum_{j=0}^{i} |\varphi(j)| K(i,j). \tag{4.102}$$

For $\varepsilon = 0$, the control problem (4.101), (4.2) coincides with the linear quadratic problem (4.1), (4.2) that has the optimal control (4.61). Let us show that the control (4.61) is the zeroth approximation to the optimal control of the control problem (4.101), (4.2), i.e., under the control u_0 the solution of the equation (4.101) is mean square stable, mean square summable and

$$0 \leq J(u_0) - V \leq C\varepsilon, \qquad V = \inf_{u \in \mathbf{U}} J(u).$$

Similarly to the Sect. 3.3 let $J(u)$ be the performance functional (4.2) for the quasilinear difference equation (4.101) and $J_0(u)$ be the performance functional (4.2) for the linear difference equation (4.1).

Suppose that the following conditions

$$K = \sup_{i \geq 0} \sum_{j=0}^{i} K(i,j) < \infty, \tag{4.103}$$

$$\hat{K} = \sup_{j \geq 0} \sum_{i=j+1}^{\infty} K(i,j) < \infty, \tag{4.104}$$

hold and note that

$$\bar{k} = \sup_{i \geq 1} K(i,i) \leq K. \tag{4.105}$$

Remark 4.8 In the particular case $K(i,j) = K(i-j)$ from (4.103)–(4.105) it follows that

$$K = \sum_{j=0}^{\infty} K(j), \quad \hat{K} = \sum_{j=1}^{\infty} K(j), \quad \bar{k} = K(0), \quad K = \hat{K} + \bar{k}.$$

4.2.1 Program Control

Here we will consider the control u_0 for the control problem (4.101), (4.2) as a program control, i.e., $u_0(i) = u_0(i, x_{0i})$, where x_{0i} is a solution of the equation (4.1) under the control u_0.

Theorem 4.3 *Let the conditions of the Lemmas* 4.4–4.9: (4.8), (4.9), (4.25)–(4.28), (4.49), (4.50), (4.55), (4.56) *and the conditions* (4.102)–(4.105) *hold. Then the control* $u_0(j)$ *defined by* (4.61) *is a zeroth approximation to the optimal control of the control problem* (4.101), (4.2) *and the solution of the equation* (4.101) *under the control* $u_0(j)$ *is mean square stable and mean square summable.*

Proof Let x^u be the solution of the equation (4.101), x_0^u be the solution of the equation (4.1) under a control $u = u(j)$. Similarly to the proof of Theorem 3.2 it is enough to prove that

$$|J(u) - J_0(u)| \leq C\varepsilon. \tag{4.106}$$

Via $x^u(0) = x_0^u(0)$ and (4.28) we have

$$\begin{aligned} |J(u) - J_0(u)| &\leq \mathbf{E}\sum_{j=1}^{\infty} |(x^u(j))'F(j)x^u(j) - (x_0^u(j))'F(j)x_0^u(j)| \\ &\leq F\sqrt{\sum_{j=1}^{\infty} \mathbf{E}|x^u(j) - x_0^u(j)|^2 \sum_{j=1}^{\infty} \mathbf{E}|x^u(j) + x_0^u(j)|^2}. \end{aligned} \tag{4.107}$$

Subtracting (4.1) with some control u from (4.101) with the same control u and using (4.102), we get

$$|y^u(i+1)| \leq \sum_{j=1}^{i} |a(i,j)||y^u(j)| + \varepsilon \sum_{j=0}^{i+1} |x^u(j)|K(i+1,j),$$

where $y^u(i) = x^u(i) - x_0^u(i)$. Squaring the both parts of the obtained inequality, calculating expectation and summing over $i = 0, 1, \ldots, \infty$, by virtue of (4.8), (4.103) and (1.31) for arbitrary $\alpha > 0$ we have

$$\begin{aligned} \sum_{i=1}^{\infty} \mathbf{E}|y^u(i)|^2 \leq \sum_{i=0}^{\infty} \Bigg[&(1+\alpha)A \sum_{j=1}^{i} |a(i,j)|\mathbf{E}|y^u(j)|^2 \\ &+ \varepsilon^2 \left(1 + \frac{1}{\alpha}\right) K \sum_{j=0}^{i+1} K(i+1,j)\mathbf{E}|x^u(j)|^2 \Bigg]. \end{aligned}$$

Note that from (4.9), (4.104), (4.105) it follows that

$$\sum_{i=0}^{\infty}\sum_{j=1}^{i}|a(i,j)|\mathbf{E}|y^u(j)|^2=\sum_{j=1}^{\infty}\sum_{i=j}^{\infty}|a(i,j)|\mathbf{E}|y^u(j)|^2$$
$$\leq\hat{A}\sum_{j=1}^{\infty}\mathbf{E}|y^u(j)|^2$$

and

$$\sum_{i=0}^{\infty}\sum_{j=0}^{i+1}K(i+1,j)\mathbf{E}|x^u(j)|^2=\sum_{i=0}^{\infty}\Bigg(\sum_{j=0}^{i}K(i+1,j)\mathbf{E}|x^u(j)|^2$$
$$+K(i+1,i+1)\mathbf{E}|x^u(i+1)|^2\Bigg)$$
$$=\sum_{j=0}^{\infty}\sum_{i=j}^{\infty}K(i+1,j)\mathbf{E}|x^u(j)|^2$$
$$+\sum_{j=0}^{\infty}K(j+1,j+1)\mathbf{E}|x^u(j+1)|^2$$
$$\leq(\hat{K}+\bar{k})\sum_{j=0}^{\infty}\mathbf{E}|x^u(j)|^2.$$

Therefore,

$$\sum_{i=1}^{\infty}\mathbf{E}|y^u(i)|^2\leq(1+\alpha)A\hat{A}\sum_{j=1}^{\infty}\mathbf{E}|y^u(j)|^2$$
$$+\varepsilon^2\left(1+\frac{1}{\alpha}\right)K(\hat{K}+\bar{k})\sum_{j=0}^{\infty}\mathbf{E}|x^u(j)|^2. \tag{4.108}$$

From (4.8), (4.9) it follows that $A\hat{A}<1$. Choosing α such that $0<\alpha<(A\hat{A})^{-1}-1$, we have $(1+\alpha)A\hat{A}<1$. So, from (4.108) we obtain

$$\sum_{i=1}^{\infty}\mathbf{E}|y^u(i)|^2\leq\varepsilon^2K(\hat{K}+\bar{k})\frac{1+\alpha^{-1}}{1-(1+\alpha)A\hat{A}}\sum_{i=0}^{\infty}\mathbf{E}|x^u(i)|^2. \tag{4.109}$$

Now via (4.106), (4.107), (4.109) it is enough to prove mean square summability of the process $x^u(i)$ under the control (4.61). Substituting the control (4.61) into the Eq. (4.101), represent (4.101) in the form

$$x^u(i+1) = \eta_2(i+1) + \sum_{j=0}^{i} a(i,j)x^u(j) + \varepsilon\Phi(i+1, x^u_{i+1}), \tag{4.110}$$

where

$$\eta_2(i+1) = \eta_1(i+1)+\sum_{j=0}^{i} b(i,j)\alpha_1(j)x^u_0(j)+\sum_{j=0}^{i} b(i,j)\sum_{k=0}^{j-1}\alpha_2(j,k)x^u_0(k), \tag{4.111}$$

and $\eta_1(i)$ is defined by (4.73).

From (4.110) via (4.102), (4.105) and small enough $\varepsilon > 0$ such that $\varepsilon\bar{k} < 1$ we have

$$(1-\varepsilon\bar{k})|x^u(i+1)| \le |\eta_2(i+1)| + \sum_{j=0}^{i}|a(i,j)||x^u(j)| + \varepsilon\sum_{j=0}^{i}|x^u(j)|K(i+1,j). \tag{4.112}$$

Squaring the both parts of (4.112) by virtue of the inequality (1.32) with arbitrary $\alpha_l > 0$, $l = 1, 2, 3$ and calculating the expectation, we obtain

$$\begin{aligned}(1-\varepsilon\bar{k})^2\mathbf{E}|x^u(i+1)|^2 &\le (1+\alpha_1+\alpha_2)\,\mathbf{E}|\eta_2(i+1)|^2 \\ &+ \left(1+\alpha_3+\frac{1}{\alpha_1}\right)A\sum_{j=0}^{i}|a(i,j)|\mathbf{E}|x^u(j)|^2 \\ &+ \left(1+\frac{1}{\alpha_2}+\frac{1}{\alpha_3}\right)K\varepsilon^2\sum_{j=0}^{i}\mathbf{E}|x^u(j)|^2K(i+1,j).\end{aligned} \tag{4.113}$$

Summing (4.113) over $i = 0, 1, \dots, \infty$, we have

$$\begin{aligned}(1-\varepsilon\bar{k})^2\sum_{i=1}^{\infty}\mathbf{E}|x^u(i)|^2 &\le (1+\alpha_1+\alpha_2)\sum_{i=0}^{\infty}\mathbf{E}|\eta_2(i+1)|^2 \\ &+ \left(1+\alpha_3+\frac{1}{\alpha_1}\right)A\sum_{i=0}^{\infty}\sum_{j=0}^{i}|a(i,j)|\mathbf{E}|x^u(j)|^2 \\ &+ \left(1+\frac{1}{\alpha_2}+\frac{1}{\alpha_3}\right)K\varepsilon^2\sum_{i=0}^{\infty}\sum_{j=0}^{i}\mathbf{E}|x^u(j)|^2K(i+1,j).\end{aligned}$$

Put $S(x^u) = \sum_{j=0}^{\infty}\mathbf{E}|x^u(j)|^2$ and note that via (4.9), (4.104)

$$\sum_{i=0}^{\infty}\sum_{j=0}^{i}|a(i,j)|\mathbf{E}|x^u(j)|^2 = \sum_{j=0}^{\infty}\sum_{i=j}^{\infty}|a(i,j)|\mathbf{E}|x^u(j)|^2 \leq \hat{A}S(x^u),$$

$$\sum_{i=0}^{\infty}\sum_{j=0}^{i}\mathbf{E}|x^u(j)|^2K(i+1,j) = \sum_{j=0}^{\infty}\sum_{i=j}^{\infty}K(i+1,j)\mathbf{E}|x^u(j)|^2 \leq \hat{K}S(x^u).$$

So, putting $\alpha_3 = \varepsilon\sqrt{(A\hat{A})^{-1}K\hat{K}}$, we obtain

$$S(x^u) \leq \mathbf{E}|\varphi_0(0)|^2 + \frac{1}{(1-\varepsilon\bar{k})^2} \times \left[(1+\alpha_1+\alpha_2)\,S(\eta_2) + \left(\left(1+\frac{1}{\alpha_1}\right)A\hat{A} + \left(1+\frac{1}{\alpha_2}\right)\varepsilon^2K\hat{K} + 2\varepsilon\sqrt{A\hat{A}K\hat{K}}\right)S(x^u)\right]. \tag{4.114}$$

Let us prove mean square summability of the sequence $\eta_2(i)$. Squaring the both parts of (4.111) via (1.30), calculating expectation and summing over $i = 0, 1, \ldots, \infty$, we have

$$\sum_{i=1}^{\infty}\mathbf{E}\eta_2^2(i) \leq 3\left[\sum_{i=1}^{\infty}\mathbf{E}\eta_1^2(i) + \sum_{i=0}^{\infty}\mathbf{E}\left(\sum_{j=0}^{i}b(i,j)\alpha_1(j)x_0^u(j)\right)^2 + \sum_{i=0}^{\infty}\mathbf{E}\left(\sum_{j=0}^{i}b(i,j)\sum_{k=0}^{j-1}\alpha_2(j,k)x_0^u(k)\right)^2\right].$$

Note that by virtue of (4.89), (4.25), (4.26) we obtain

$$\sum_{i=0}^{\infty}\mathbf{E}\left(\sum_{j=0}^{i}b(i,j)\alpha_1(j)x_0^u(j)\right)^2 \leq \sup_{j\geq 0}|\alpha_1(j)|^2\sum_{i=0}^{\infty}\sum_{l=0}^{i}|b(i,l)|\sum_{j=0}^{i}|b(i,j)|\mathbf{E}|x_0^u(j)|^2$$

$$\leq \sup_{j\geq 0}|\alpha_1(j)|^2\sup_{i\geq 0}\sum_{l=0}^{i}|b(i,l)|\sum_{j=0}^{\infty}\sum_{i=j}^{\infty}|b(i,j)|\mathbf{E}|x_0^u(j)|^2$$
$$\leq \hat{\Psi}^2(1+R)^2B\hat{B}S(x^u)$$

and similarly via (4.91)

$$\sum_{i=0}^{\infty}\mathbf{E}\left(\sum_{j=0}^{i}b(i,j)\sum_{k=0}^{j-1}\alpha_2(j,k)x_0^u(k)\right)^2$$
$$\leq \sup_{j\geq 0}\left|\sum_{k=0}^{j-1}\alpha_2(j,k)\right|^2\sum_{i=0}^{\infty}\sum_{l=0}^{i}|b(i,l)|\sum_{j=0}^{i}|b(i,j)|\mathbf{E}|x_0^u(j)|^2$$
$$\leq \hat{\Psi}^2(1+R)^2(2+3R_1)^2B\hat{B}S(x^u).$$

So,

$$S(\eta_2)\leq \mathbf{E}|\eta_2(0)|^2+3\left[S(\eta_1)+\hat{\Psi}^2(1+R)^2(1+(2+3R_1)^2)B\hat{B}S(x_0^u)\right]. \quad (4.115)$$

From Lemma 4.8 it follows that from mean square summability of the sequence $\eta(i)$ mean square summability of the sequence $\eta_1(i)$ follows. From Theorem 4.2, it follows that the sequence $x_0^u(i)$ is mean square summable too. Therefore, from the inequality (4.115) it follows that the sequence $\eta_2(i)$ is mean square summable.

From (4.114) it follows that

$$\left[1-\frac{1}{(1-\varepsilon\bar{k})^2}\left(\left(1+\frac{1}{\alpha_1}\right)A\hat{A}+\left(1+\frac{1}{\alpha_2}\right)\varepsilon^2K\hat{K}+2\varepsilon\sqrt{A\hat{A}K\hat{K}}\right)\right]S(x^u)$$
$$\leq \mathbf{E}|\varphi_0(0)|^2+\frac{1+\alpha_1+\alpha_2}{(1-\varepsilon\bar{k})^2}S(\eta_2). \quad (4.116)$$

So, by the conditions (4.8), (4.9), (4.103)–(4.105) for

$$0<\varepsilon<\frac{1-\sqrt{A\hat{A}}}{\bar{k}+\sqrt{K\hat{K}}} \quad (4.117)$$

there exist big enough $\alpha_1>0$, $\alpha_2>0$ such that the condition

$$\frac{1}{(1-\varepsilon\bar{k})^2}\left[\left(1+\frac{1}{\alpha_1}\right)A\hat{A}+\left(1+\frac{1}{\alpha_2}\right)\varepsilon^2K\hat{K}+2\varepsilon\sqrt{A\hat{A}K\hat{K}}\right]<1 \quad (4.118)$$

holds. From this and (4.116) it follows that the solution x^u of the Eq. (4.101) is mean square summable. Therefore, via (4.107), (4.109) the condition (4.106) holds.

Let us prove now mean square stability of the solution of the equation (4.101) under the control u_0 given in (4.61). Putting in (4.113) $u = u_0$ and $\alpha_3 = \varepsilon$, rewrite this inequality in the form

$$y(i+1) \le \tilde{\eta}(i+1) + \sum_{j=0}^{i} P(i,j)y(j), \tag{4.119}$$

where

$$\begin{aligned} y(i) &= \mathbf{E}|x^{u_0}(i)|^2, \\ \tilde{\eta}(i) &= \frac{1}{(1-\varepsilon\bar{k})^2}(1+\alpha_1+\alpha_2)\,\mathbf{E}|\eta_2(i)|^2, \end{aligned} \tag{4.120}$$

$$\begin{aligned} P(i,j) = \frac{1}{(1-\varepsilon\bar{k})^2}\Bigg[\left(1+\frac{1}{\alpha_1}+\varepsilon\right)A|a(i,j)| \\ + \left(1+\left(1+\frac{1}{\alpha_2}\right)\varepsilon\right)\varepsilon KK(i+1,j)\Bigg]. \end{aligned} \tag{4.121}$$

From (4.115) it follows, in particular, that the sequence $\eta_2(i)$ is uniformly mean square bounded. From (4.120) it follows that the sequence $\tilde{\eta}(i)$ is uniformly bounded. Via Lemma 1.5 from (4.119) it follows that $y(i) \le z(i)$, $i = 0, 1, \ldots$, where $z(i)$ is the solution of the equation

$$z(i+1) = \tilde{\eta}(i+1) + \sum_{j=0}^{i} P(i,j)z(j). \tag{4.122}$$

Besides from the conditions (4.8), (4.103) and (4.121) it follows that there exist big enough $\alpha_1 > 0$ and small enough $\varepsilon > 0$ such that the condition

$$\sup_{i\ge 0}\sum_{j=0}^{i} P(i,j) < 1$$

holds. Via Lemma 4.2 there exists the resolvent $R(i,j)$ of the kernel $P(i,j)$ that satisfies the condition (4.4). So, via (4.122), (4.120) for some $C > 0$ we obtain

$$\begin{aligned} y(i+1) &\le z(i+1) \\ &= \tilde{\eta}(i+1) + \sum_{j=0}^{i} R(i,j)\tilde{\eta}(j) \\ &\le C\|\eta_2\|^2 \end{aligned}$$

or $\|x^{u_0}\|^2 \le C\|\eta_2\|^2$. From the condition (4.111) via the conditions (4.25), (4.77), (4.89), (4.91) it follows that

$$\|x^{u_0}\|^2 \le C(\|\eta\|^2 + \|x_0^{u_0}\|^2),$$

where $x_0^{u_0}$ is the solution of the equation (4.1) under the control (4.61). Via Theorem 4.2 the solution of the equation (4.1) under the control (4.61) is mean square stable and therefore there exists some $C > 0$ such that $\|x_0^{u_0}\|^2 < C\|\eta\|^2$. So, as a result we obtain $\|x^{u_0}\|^2 \le C\|\eta\|^2$ that means that the solution of the equation (4.101) under the control u_0 is mean square stable. The theorem is proved.

4.2.2 Feedback Control

Let us prove now that the control $\hat{u}_0(j, x_j)$ defined by (4.61) is a zeroth approximation to the optimal control of the control problem (4.101), (4.2) if it is considered as a feedback control too.

Theorem 4.4 *Let the conditions of Theorem 4.3 and the condition* (4.33) *hold. Then the control* $\hat{u}_0(j, x_j)$ *defined by* (4.61) *is a zeroth approximation to the optimal control of the control problem* (4.101), (4.2) *and the solution of the equation* (4.101) *under the control* $\hat{u}_0(j, x_j)$ *is mean square stable and mean square summable.*

Proof Let $J(\hat{u}_0)$ be the performance functional (4.2) for the quasilinear equation (4.101) and $J_0(\hat{u}_0)$ be the performance functional (4.2) for the linear equation (4.1) under the control $\hat{u}_0$. Similarly to the proof of Theorem 4.3 it is enough to prove that

$$|J(\hat{u}_0) - J_0(\hat{u}_0)| \le C\varepsilon. \tag{4.123}$$

Let $\hat{x}_0$ be the solution of the equation (4.101) and x_0 be the solution of the equation (4.1) under the control $\hat{u}_0$. From (4.2), using (4.28), (4.33) and $\hat{x}_{00} = x_{00}$, similarly to (4.107) we have

$$\begin{aligned}
&|J(\hat{u}_0) - J_0(\hat{u}_0)| \\
&\le \mathbf{E}\sum_{j=0}^{\infty}[\hat{x}_0'(j)F(j)\hat{x}_0(j) - x_0'(j)F(j)x_0(j)] \\
&\quad + \mathbf{E}\sum_{j=0}^{\infty}[\hat{u}_0'(j, \hat{x}_{0j})G(j)\hat{u}_0(j, \hat{x}_{0j}) - \hat{u}_0'(j, x_{0j})G(j)\hat{u}_0(j, x_{0j})]
\end{aligned}$$

$$\leq F\sqrt{\sum_{j=1}^{\infty}\mathbf{E}|\hat{x}_0(j)-x_0(j)|^2\sum_{j=1}^{\infty}\mathbf{E}|\hat{x}_0(j)+x_0(j)|^2}$$

$$+G\sqrt{\sum_{j=1}^{\infty}\mathbf{E}|\hat{u}_0(j,\hat{x}_{0j})-\hat{u}_0(j,x_{0j})|^2\sum_{j=1}^{\infty}\mathbf{E}|\hat{u}_0(j,\hat{x}_{0j})+\hat{u}_0(j,x_{0j})|^2}.$$

Subtracting (4.1) from (4.101) and using (4.74), (4.102), for $y(i)=\hat{x}_0(i)-x_0(i)$ we obtain

$$|y(i+1)|\leq\sum_{j=1}^{i}|a_1(i,j)||y(j)|+\varepsilon\sum_{j=0}^{i+1}|\hat{x}_0(j)|K(i+1,j). \tag{4.124}$$

Squaring the both parts of (4.124), calculating expectation and summing over $i=0,\ldots,\infty$, by virtue of (4.86), (4.88), (4.104), (4.105), similarly to (4.109), we obtain by $(1+\alpha)A_1\hat{A}_1<1$

$$\sum_{i=1}^{\infty}\mathbf{E}|y(i)|^2\leq\varepsilon^2K(\hat{K}+\bar{k})\frac{1+\alpha^{-1}}{1-(1+\alpha)A_1\hat{A}_1}\sum_{i=0}^{\infty}\mathbf{E}|\hat{x}_0(i)|^2. \tag{4.125}$$

From (4.61) it follows that

$$|\hat{u}_0(j+1,\hat{x}_{0,j+1})-\hat{u}_0(j+1,x_{0,j+1})|$$
$$\leq|\alpha_1(j+1)||y(j+1)|+\sum_{k=1}^{j}|\alpha_2(j+1,k)||y(k)|.$$

From this via (4.89), (4.91) we have

$$|\hat{u}_0(j+1,\hat{x}_{0,j+1})-\hat{u}_0(j+1,x_{0,j+1})|^2$$
$$\leq 2\left[|\alpha_1(j+1)|^2|y(j+1)|^2+\sum_{l=1}^{j}|\alpha_2(j+1,l)|\sum_{k=1}^{j}|\alpha_2(j+1,k)||y(k)|^2\right]$$
$$\leq 2\left[\hat{\Psi}^2(1+R)^2|y(j+1)|^2+\hat{\Psi}(1+R)(2+3R_1)\sum_{k=1}^{j}|\alpha_2(j+1,k)||y(k)|^2\right]$$
$$\leq 2\hat{\Psi}(1+R)\left[\hat{\Psi}(1+R)|y(j+1)|^2+(2+3R_1)\sum_{k=1}^{j}|\alpha_2(j+1,k)||y(k)|^2\right].$$

Calculating expectation and summing the obtained inequality over $j=0,1,\ldots,\infty$, we obtain

$$\sum_{j=1}^{\infty} \mathbf{E}|\hat{u}_0(j,\hat{x}_{0,j}) - \hat{u}_0(j,x_{0,j})|^2$$
$$\leq 2\hat{\Psi}(1+R)\Bigg[\hat{\Psi}(1+R)\sum_{j=1}^{\infty}\mathbf{E}|y(j)|^2$$
$$+ (2+3R_1)\sum_{j=1}^{\infty}\sum_{k=1}^{j}|\alpha_2(j+1,k)|\mathbf{E}|y(k)|^2\Bigg].$$

Via (4.100) we have

$$\sum_{j=1}^{\infty}\sum_{k=1}^{j}|\alpha_2(j+1,k)|\mathbf{E}|y(k)|^2$$
$$= \sum_{k=1}^{\infty}\sum_{j=k}^{\infty}|\alpha_2(j+1,k)|\mathbf{E}|y(k)|^2$$
$$\leq \left(\hat{\Psi}(1+R)\hat{R}_1 + \left(1+\hat{R}_1\right)\hat{A}\left(\hat{\Psi}(1+R) + \Psi\left(1+\hat{R}\right)\right)\right)\sum_{k=1}^{\infty}\mathbf{E}|y(k)|^2.$$

So,

$$\sum_{j=1}^{\infty} \mathbf{E}|\hat{u}_0(j,\hat{x}_{0,j}) - \hat{u}_0(j,x_{0,j})|^2$$
$$\leq 2\hat{\Psi}(1+R)\Bigg[\hat{\Psi}(1+R) + (2+3R_1)\bigg(\hat{\Psi}(1+R)\hat{R}_1 + \left(1+\hat{R}_1\right)\hat{A}$$
$$\times\left(\hat{\Psi}(1+R) + \Psi\left(1+\hat{R}\right)\right)\bigg)\Bigg]S(y). \tag{4.126}$$

Besides note that from (4.61) we have

$$\mathbf{E}|\hat{u}_0(0,\hat{x}_{00})|^2 \leq 2\left[\mathbf{E}|\alpha_0(0)|^2 + |\alpha_1(0)|^2\mathbf{E}|x_0(0)|^2\right], \tag{4.127}$$

$$\mathbf{E}|\hat{u}_0(j+1,\hat{x}_{0,j+1})|^2 \leq 3\Bigg[\mathbf{E}|\alpha_0(j+1)|^2 + |\alpha_1(j+1)|^2\mathbf{E}|x_0(j+1)|^2$$
$$+ \sum_{l=0}^{j}\alpha_2(j+1,l)\sum_{k=0}^{j}\alpha_2(j+1,k)\mathbf{E}|x_0(k)|^2\Bigg]. \tag{4.128}$$

Summing the inequality (4.128) over $j = 0, 1, \ldots, \infty$ together with (4.127), via (4.89), (4.91) and (4.79), (4.84), (4.93), (4.99) we obtain

$$\sum_{j=0}^{\infty}\mathbf{E}|\hat{u}_0(j,\hat{x}_{0,j})|^2$$
$$\leq 3\Bigg[\sum_{j=0}^{\infty}\mathbf{E}|\alpha_0(j)|^2+\sum_{j=0}^{\infty}|\alpha_1(j)|^2\mathbf{E}|\hat{x}_0(j)|^2$$
$$+\sum_{j=0}^{\infty}\sum_{l=0}^{j}\alpha_2(j+1,l)\sum_{k=0}^{j}\alpha_2(j+1,k)\mathbf{E}|\hat{x}_0(k)|^2\Bigg]$$
$$\leq 3\Bigg[S(\alpha_0)+\hat{\Psi}^2(1+R)^2S(\hat{x}_0)$$
$$+\hat{\Psi}(1+R)(2+3R_1)\sum_{k=0}^{\infty}\sum_{j=k}^{\infty}\alpha_2(j+1,k)\mathbf{E}|\hat{x}_0(k)|^2\Bigg]$$
$$\leq 3\Bigg[\hat{\Psi}\left(1+\sqrt{R_1\hat{R}_1}\right)^2\left((1+\sqrt{R\hat{R}})\sqrt{\Psi}+(1+R)\sqrt{\hat{\Psi}}\right)^2S(\eta)$$
$$+\hat{\Psi}(1+R)\Bigg(\hat{\Psi}(1+R)+(2+3R_1)\Bigg(\hat{\Psi}(1+R)\hat{R}_1$$
$$+\left(1+\hat{R}_1\right)\hat{A}\left[\hat{\Psi}(1+R)+\Psi\left(1+\hat{R}\right)\right]\Bigg)\Bigg)S(\hat{x}_0)\Bigg].$$

So, similarly to the proof of Theorem 4.3 now it is necessary to prove mean square summability of the process $\hat{x}_0$. Substituting (4.61) into (4.101) and using (4.73), (4.74), rewrite the Eq. (4.95) in the form

$$\hat{x}_0(i+1)=\eta_1(i+1)+\sum_{j=0}^{i}a_1(i,j)\hat{x}_0(j)+\varepsilon\Phi(i+1,\hat{x}_{0,i+1}).$$

From this via (4.102), (4.105) and small enough $\varepsilon>0$ such that $\varepsilon\bar{k}<1$ we have

$$(1-\varepsilon\bar{k})|\hat{x}_0(i+1)|\leq|\eta_1(i+1)|+\sum_{j=0}^{i}|a_1(i,j)||\hat{x}_0(j)|+\varepsilon\sum_{j=0}^{i}|x^u(j)|K(i+1,j).\tag{4.129}$$

Squaring the both parts of (4.129) by virtue of the inequality (1.32) with arbitrary $\alpha_l>0$, $l=1,2,3$ and calculating the expectation, similarly to (4.113) we obtain

$$(1-\varepsilon\bar{k})^2\mathbf{E}|\hat{x}_0(i+1)|^2\leq(1+\alpha_1+\alpha_2)\,\mathbf{E}|\eta_1(i+1)|^2$$
$$+\left(1+\alpha_3+\frac{1}{\alpha_1}\right)A_1\sum_{j=0}^{i}|a_1(i,j)|\mathbf{E}|\hat{x}_0(j)|^2$$

$$+\left(1+\frac{1}{\alpha_2}+\frac{1}{\alpha_3}\right)K\varepsilon^2\sum_{j=0}^{i}\mathbf{E}|\hat{x}_0(j)|^2K(i+1,j). \tag{4.130}$$

Summing (4.130) over $i=0,1,\ldots,\infty$, similarly to (4.113) we have

$$\begin{aligned}(1-\varepsilon\bar{k})^2\sum_{i=1}^{\infty}\mathbf{E}|\hat{x}_0(i)|^2 &\le (1+\alpha_1+\alpha_2)\sum_{i=0}^{\infty}\mathbf{E}|\eta_1(i+1)|^2\\ &+\left(1+\alpha_3+\frac{1}{\alpha_1}\right)A_1\sum_{i=0}^{\infty}\sum_{j=0}^{i}|a_1(i,j)|\mathbf{E}|\hat{x}_0(j)|^2\\ &+\left(1+\frac{1}{\alpha_2}+\frac{1}{\alpha_3}\right)K\varepsilon^2\sum_{i=0}^{\infty}\sum_{j=0}^{i}\mathbf{E}|\hat{x}_0(j)|^2K(i+1,j)\end{aligned}$$

and via (4.88), (4.104)

$$\begin{aligned}\sum_{i=0}^{\infty}\sum_{j=0}^{i}|a_1(i,j)|\mathbf{E}|\hat{x}_0(j)|^2 &= \sum_{j=0}^{\infty}\sum_{i=j}^{\infty}|a_1(i,j)|\mathbf{E}|\hat{x}_0(j)|^2\\ &\le \hat{A}_1S(\hat{x}_0),\\ \sum_{i=0}^{\infty}\sum_{j=0}^{i}\mathbf{E}|\hat{x}_0(j)|^2K(i+1,j) &= \sum_{j=0}^{\infty}\sum_{i=j}^{\infty}K(i+1,j)\mathbf{E}|\hat{x}_0(j)|^2\\ &\le \hat{K}S(\hat{x}_0).\end{aligned}$$

So, putting $\alpha_3=\varepsilon\sqrt{(A_1\hat{A}_1)^{-1}K\hat{K}}$, we obtain

$$\begin{aligned}S(\hat{x}_0) \le \mathbf{E}|\varphi_0(0)|^2 &+ \frac{1}{(1-\varepsilon\bar{k})^2}\Bigg[(1+\alpha_1+\alpha_2)S(\eta_1)\\ &+\left(\left(1+\frac{1}{\alpha_1}\right)A_1\hat{A}_1+\left(1+\frac{1}{\alpha_2}\right)\varepsilon^2K\hat{K}\right.\\ &\left.+2\varepsilon\sqrt{A_1\hat{A}_1K\hat{K}}\right)S(\hat{x}_0)\Bigg].\end{aligned}$$

From this similarly to the condition (4.116) it follows that

$$\begin{aligned}&\left[1-\frac{1}{(1-\varepsilon\bar{k})^2}\left(\left(1+\frac{1}{\alpha_1}\right)A_1\hat{A}_1+\left(1+\frac{1}{\alpha_2}\right)\varepsilon^2K\hat{K}+2\varepsilon\sqrt{A_1\hat{A}_1K\hat{K}}\right)\right]S(\hat{x}_0)\\ &\quad\le \mathbf{E}|\varphi_0(0)|^2+\frac{1+\alpha_1+\alpha_2}{(1-\varepsilon\bar{k})^2}S(\eta_1)\end{aligned} \tag{4.131}$$

for small enough $\varepsilon > 0$ and big enough $\alpha_1 > 0$ and $\alpha_2 > 0$ such that

$$0 < \varepsilon < \frac{1-\sqrt{A_1\hat{A}_1}}{\bar{k}+\sqrt{K\hat{K}}},$$

$$\frac{1}{(1-\varepsilon\bar{k})^2}\left(\left(1+\frac{1}{\alpha_1}\right)A_1\hat{A}_1+\left(1+\frac{1}{\alpha_2}\right)\varepsilon^2K\hat{K}+2\varepsilon\sqrt{A_1\hat{A}_1K\hat{K}}\right) < 1. \quad (4.132)$$

Via Lemma 4.8 if the sequence $\eta(i)$ is mean square summable then the sequence $\eta_1(i)$ is mean square summable too. So, from (4.131), (4.132) it follows that the solution of the equation (4.101) under the control $\hat{u}_0(j, x_j)$ is mean square summable.

Let us prove mean square stability of the solution of the equation (4.101) under the control $\hat{u}_0(j, x_j)$. Putting in (4.130) $\alpha_3 = \varepsilon$, represent (4.130) in the form

$$y(i+1) \leq \tilde{\eta}(i+1) + \sum_{j=0}^{i} P_1(i,j)y(j), \quad (4.133)$$

where

$$y(i) = \mathbf{E}|\hat{x}_0(i)|^2,$$

$$\tilde{\eta}(i) = \frac{1+\alpha_1+\alpha_2}{(1-\varepsilon\bar{k})^2}\mathbf{E}|\eta_1(i)|^2, \quad (4.134)$$

$$P_1(i,j) = \frac{1}{(1-\varepsilon\bar{k})^2}\left[\left(1+\frac{1}{\alpha_1}+\varepsilon\right)A_1|a_1(i,j)| + \left(1+\left(1+\frac{1}{\alpha_2}\right)\varepsilon\right)\varepsilon KK(i+1,j)\right]. \quad (4.135)$$

Via Lemma 1.5 from (4.133) it follows that $y(i) \leq z(i)$, $i = 0, 1, \ldots$, where $z(i)$ is the solution of the equation

$$z(i+1) = \tilde{\eta}(i+1) + \sum_{j=0}^{i} P(i,j)z(j). \quad (4.136)$$

Besides from the conditions (4.86), (4.103) and (4.135) it follows that there exist big enough $\alpha_1 > 0$ and small enough $\varepsilon > 0$ such that the condition

$$\sup_{i\geq 0}\sum_{j=0}^{i} P(i,j) < 1$$

holds. Via Lemma 4.2 there exists the resolvent $R(i, j)$ of the kernel $P_1(i, j)$ that satisfies the condition (4.4). So, via (4.136), (4.134) for some $C > 0$ we obtain

$$\begin{aligned} y(i+1) &\le z(i+1) \\ &= \tilde{\eta}(i+1) + \sum_{j=0}^{i} R(i,j)\tilde{\eta}(j) \\ &\le C\|\eta_1\|^2. \end{aligned}$$

From this and (4.77) it follows that there exists some $C > 0$ such that $\|\hat{x}_0\|^2 < C\|\eta\|^2$, i.e., the solution of the equation (4.101) under the control $\hat{u}_0(j, x_j)$ is mean square stable. The theorem is proven.

Chapter 5
Optimal Estimation

In this chapter, we consider the problem of constructing the optimal (in the mean square sense) estimate of an arbitrary partially observable Gaussian stochastic process from its observations with delay. It is proved that the desired estimate is defined by a unique solution of the fundamental filtering equation of the Wiener–Hopf type. A qualitative analysis of this equation is made, and several cases where it can be solved analytically are considered. The relationship between the observation error and the magnitude of delay in observations is investigated. It is shown that the fundamental filtering equation also describes the solutions of the forecasting and interpolation problems. In the case if the unobservable process is given by a stochastic difference Volterra equation, an analogue of the Kalman–Bucy filter is constructed: the system of four stochastic difference equations defines the optimal in the mean square sense estimate.

5.1 Filtering Problem

5.1.1 Fundamental Filtering Equation

In this section, it is shown that the solution of the filtering problem can be reduced to the solution of the difference Wiener–Hopf equation called the fundamental filtering equation. It is proved that the solution of the filtering problem is a solution of some auxiliary (dual) problem of the optimal control. Some particular cases of the fundamental filtering equation are considered in which its solution can be obtained in a final form. A dependence of the estimation error on a magnitude of delay in the observations is investigated.

Let $\{\Omega, \mathfrak{F}, \mathbf{P}\}$ be a basic probability space with a family of σ-algebras $\mathfrak{F}_i \subset \mathfrak{S}, i \in Z = \{0, 1, \ldots, N\}$. Let $(x(i), y(i))$ be a partially observable $\mathfrak{F}_i$-adapted stochastic process, $x(i) \in \mathbf{R}^n$ and $y(i) \in \mathbf{R}^k$ are unobservable and observable components, respectively.

L. Shaikhet, *Optimal Control of Stochastic Difference Volterra Equations*, Studies in Systems, Decision and Control 17, DOI 10.1007/978-3-319-13239-6_5

Consider the problem of constructing the optimal (in the mean square sense) estimate $m_0(N)$ of the random variable $x(N)$ from the results of observations of $y(i)$, $1 \leq i \leq N$. If $\mathbf{E}|x(N)|^2 < \infty$ then the desired estimate is the conditional expectation [130]

$$m_0(N) = \mathbf{E}(x(N)/\mathfrak{F}_N^y). \tag{5.1}$$

Here $\mathfrak{F}_i^y$ is the minimal σ-algebra generated by the process $y(j), 0 \leq j \leq i$.

Let $x(i)$ be a Gaussian stochastic process and $y(i)$ is defined by the relation

$$\begin{aligned} y(i+1) &= A(i)x(i-h) + \xi(i+1), \\ y(0) &= 0, \quad i = 0, 1, \ldots, N-1, \\ x(j) &= 0, \quad -h \leq j < 0, \end{aligned} \tag{5.2}$$

where $A(i)$ is a nonrandom matrix of the dimension $k \times n$, $\xi(i) \in \mathbf{R}^k$ are $\mathfrak{F}_i$-adapted independent from each other Gaussian random variables, such that

$$\begin{aligned} &\mathbf{E}x(i) = 0, \qquad \mathbf{E}x(i)x'(j) = R(i, j), \\ &\mathbf{E}\xi(i) = 0, \qquad \mathbf{E}\xi(i)\xi'(j) = 0, \quad i \neq j, \qquad \mathbf{E}\xi(i)\xi'(i) = S(i), \\ &\mathbf{E}x(i)\xi'(j) = Q(i, j), \qquad Q(i, j) = 0 \quad \text{for } j > i, \end{aligned} \tag{5.3}$$

the matrices $S(i)$, $R(i, j)$, and $Q(i, j)$ have appropriate dimensions and the matrix $S(i)$ is uniformly positive definite.

Lemma 5.1 *There exists a nonrandom matrix $u_0(j)$, $j \in Z$, of the dimension $n \times k$ such that the estimate* (5.1) *has the representation*

$$m_0(N) = \sum_{j=0}^{N-1} u_0(j)y(j+1). \tag{5.4}$$

The proof of this lemma follows from the Theorem 1.1 (about normal correlation, [130]) and Corollary 1.2.

From Lemma 5.1 it follows that the construction of optimal (in the mean square sense) estimate (5.1) is reduced to the construction of a certain nonrandom matrix $u_0(j)$ for the representation (5.4).

Theorem 5.1 *The matrix $u_0(j)$ defining the estimate* (5.4) *is a unique solution of the equation*

$$u_0(j)S(j+1) + \sum_{i=0}^{N-1} u_0(i)Z(i, j) = P(j), \quad j = 0, 1, \ldots, N-1, \tag{5.5}$$

where

$$\begin{aligned} P(j) &= R(N, j-h)A'(j) + Q(N, j+1), \\ Z(i, j) &= A(i)R(i-h, j-h)A'(j) + A(i)Q(i-h, j+1) \\ &\quad + Q'(j-h, i+1)A'(j). \end{aligned} \tag{5.6}$$

Proof Let $m_0(N)$ be the optimal (in the man square sense) estimate of the variable $x(N)$ defined by the relation (5.4) and let $m(N)$ be an estimate of the form (5.4) with an arbitrary matrix $u(j)$. Then via (5.1) we have

$$\begin{aligned} \mathbf{E}(x(N) - m_0(N))m'(N) &= \mathbf{E}[\mathbf{E}(x(N) - m_0(N))/\mathfrak{F}_N^y]m'(N) \\ &= \mathbf{E}[\mathbf{E}(x(N)/\mathfrak{F}_N^y) - m_0(N)]m'(N) \\ &= 0. \end{aligned} \tag{5.7}$$

Using for the estimates $m_0(N)$ and $m(N)$ representations of the form (5.4), rewrite (5.7) in the form

$$\sum_{j=0}^{N-1}\left(\mathbf{E}x(N)y'(j+1) - \sum_{i=0}^{N-1} u_0(i)\mathbf{E}y(i+1)y'(j+1)\right)u'(j) = 0. \tag{5.8}$$

Since $u(j)$ is an arbitrary matrix, the expression in the brackets in (5.8) must be zero for each $j = 0, \ldots, N-1$. So, the equation for the matrix $u_0(i)$ has the form

$$\mathbf{E}x(N)y'(j+1) = \sum_{i=0}^{N-1} u_0(i)\mathbf{E}y(i+1)y'(j+1). \tag{5.9}$$

Using (5.2), (5.3), (5.6), we get

$$\begin{aligned} \mathbf{E}x(N)y'(j+1) &= \mathbf{E}x(N)x'(j-h)A'(j) + \mathbf{E}x(N)\xi'(j+1) \\ &= R(N, j-h)A'(j) + Q(N, j+1) \\ &= P(j), \end{aligned} \tag{5.10}$$

and

$$\begin{aligned} &\sum_{i=0}^{N-1} u_0(i)\mathbf{E}y(i+1)y'(j+1) \\ &\quad = \sum_{i=0}^{N-1} u_0(i)\mathbf{E}(A(i)x(i-h) + \xi(i+1))(A(j)x(j-h) + \xi(j+1))' \end{aligned}$$

$$
\begin{aligned}
&= \sum_{i=0}^{N-1} u_0(i)\Big[A(i)R(i-h, j-h)A'(j) + Q'(j-h, i+1)A'(j) \\
&\qquad\qquad + A(i)Q(i-h, j+1)\Big] + u_0(j)S(j+1) \\
&= \sum_{i=0}^{N-1} u_0(i)Z(i, j) + u_0(j)S(j+1). \qquad (5.11)
\end{aligned}
$$

The relations (5.9)–(5.11) imply (5.5) and (5.6). Thus, the function $u_0(i)$ defining the optimal (in the mean square sense) estimate (5.4), is a solution of the equation (5.5). This means, in particular, that a solution of the equation (5.5) exists.

Let us prove now that this solution is unique. Let us suppose that there are two different solutions of the equation (5.5): $u_1(j)$ and $u_2(j)$. Put $\Delta u(j) = u_1(j) - u_2(j)$. Substituting $u_1(j)$ and $u_2(j)$ into (5.9) and subtracting the one equality from the second one, we obtain

$$
\sum_{i=0}^{N-1} \Delta u(i)\mathbf{E}y(i+1)y'(j+1) = 0. \qquad (5.12)
$$

Multiplying (5.12) from the right by $\Delta u'(i)$ and summing over $j = 0, \ldots, N-1$, we get the zero matrix

$$
\sum_{j=0}^{N-1}\sum_{i=0}^{N-1} \Delta u(i)\mathbf{E}y(i+1)y'(j+1)\Delta u'(j) = 0.
$$

Calculating the trace of this matrix, we obtain

$$
\mathbf{E}\left|\sum_{i=0}^{N-1} \Delta u(i)y(i+1)\right|^2 = 0. \qquad (5.13)
$$

Put now

$$
x_0(i) = \sum_{j=0}^{i-1} \alpha(i, j+1)\xi(j+1), \qquad \alpha(i, j) = Q(i, j)S^{-1}(j). \qquad (5.14)
$$

Via (5.3), (5.14) we have

$$
\begin{aligned}
\mathbf{E}[x(i) - x_0(i)]\xi'(j+1) &= Q(i, j+1) - \mathbf{E}\sum_{l=0}^{N-1} \alpha(i, l+1)\xi(l+1)\xi'(j+1) \\
&= Q(i, j+1) - \alpha(i, j+1)S(j+1) \\
&= 0. \qquad (5.15)
\end{aligned}
$$

So, using (5.2), (5.14), represent (5.13) in the form

$$\begin{aligned}
\mathbf{E}\left|\sum_{i=0}^{N-1}\Delta u(i)y(i+1)\right|^2 &= \mathbf{E}\left|\sum_{i=0}^{N-1}\Delta u(i)(A(i)x(i-h)+\xi(i+1))\right|^2 \\
&= \mathbf{E}\left|\sum_{i=0}^{N-1}\Delta u(i)\Big(A(i)(x(i-h)-x_0(i-h))\right. \\
&\qquad \left. + A(i)x_0(i-h)+\xi(i+1)\Big)\right|^2 \\
&= \mathbf{E}\left|\sum_{i=0}^{N-1}\Delta u(i)A(i)(x(i-h)-x_0(i-h))\right. \\
&\qquad + \sum_{i=0}^{N-1}\Delta u(i)A(i)\sum_{j=0}^{i-h-1}\alpha(i-h,j+1)\xi(j+1) \\
&\qquad \left. + \sum_{i=0}^{N-1}\Delta u(i)\xi(i+1)\right|^2. \qquad (5.16)
\end{aligned}$$

Transform two last sums by the following way

$$\begin{aligned}
&\sum_{i=0}^{N-1}\Delta u(i)A(i)\sum_{j=0}^{i-h-1}\alpha(i-h,j+1)\xi(j+1)+\sum_{i=0}^{N-1}\Delta u(i)\xi(i+1) \\
&\quad = \sum_{j=0}^{N-h-2}\sum_{i=j+h+1}^{N-1}\Delta u(i)A(i)\alpha(i-h,j+1)\xi(j+1)+\sum_{j=0}^{N-1}\Delta u(j)\xi(j+1) \\
&\quad = \sum_{j=0}^{N-1}\beta(j)\xi(j+1), \qquad (5.17)
\end{aligned}$$

where

$$\beta(j)=\begin{cases}\Delta u(j)+\sum\limits_{i=j+h+1}^{N-1}\Delta u(i)A(i)\alpha(i-h,j+1), & 0\le j\le N-h-2,\\ \Delta u(j), & N-h-1\le j\le N-1.\end{cases} \qquad (5.18)$$

As a result, using (5.15) and the properties (5.3) of the process $\xi(j)$, from (5.13), (5.16), (5.17), we obtain

$$
\begin{aligned}
\mathbf{E}\left|\sum_{i=0}^{N-1}\Delta u(i)y(i+1)\right|^2 &= \mathbf{E}\left|\sum_{i=0}^{N-1}\Delta u(i)A(i)(x(i-h)-x_0(i-h))\right. \\
&\quad \left. + \sum_{j=0}^{N-1}\beta(j)\xi(j+1)\right|^2 \\
&= \mathbf{E}\left|\sum_{i=0}^{N-1}\Delta u(i)A(i)(x(i-h)-x_0(i-h))\right|^2 \\
&\quad + \sum_{j=0}^{N-1}Tr\big[\beta(j)S(j+1)\beta'(j)\big] \\
&= 0.
\end{aligned}
$$

Since the matrix $S(j)$ is positive definite, each from this summands equals zero and therefore for each $0 \le j \le N-1$

$$
Tr\big[\beta(j)S(j+1)\beta'(j)\big] = \sum_{l=1}^{n}\big[\beta_l(j)S(j+1)\beta_l'(j)\big] = 0,
$$

where $\beta_l(j)$, $1 \le l \le n$, are the lines of the matrix $\beta(j)$. From this and a positive definiteness of the matrix $S(j)$ it follows also that $\beta_l(j) = 0$, $1 \le l \le n$. So, $\beta(j) = 0$ and via (5.18), we get the equation for $\Delta u(j)$:

$$
\Delta u(j) = 0, \quad N-h-1 \le j \le N-1, \tag{5.19}
$$

and

$$
\Delta u(j) + \sum_{i=j+h+1}^{N-1}\Delta u(i)A(i)\alpha(i-h, j+1) = 0, \quad 0 \le j \le N-h-2. \tag{5.20}
$$

Using (5.19), let us suppose that

$$
\Delta u(j) = 0 \quad \text{for } N-h-l \le j \le N-1, \quad l \ge 1, \tag{5.21}
$$

and prove that $\Delta u(N-h-l-1) = 0$, $l \ge 1$. Really, for $j = N-h-l-1$ from (5.20) we have

$$
\Delta u(N-h-l-1) + \sum_{i=N-l}^{N-1}\Delta u(i)A(i)\alpha(i-h, N-h-l) = 0.
$$

Since $N-l \ge N-h-l$ then via (5.21) $\Delta u(i) = 0$ for $N-l \le i \le N-1$ and therefore $\Delta u(N-h-l-1) = 0$, $l \ge 1$. It means that $\Delta u(j) = 0$ for all

$j = 0, 1, \ldots, N-1$. Thus, the both solutions $u_1(j)$ and $u_2(j)$ coincide, i.e., the solution of the considered filtering problem is unique. The proof is completed.

5.1.2 Dual Problem of Optimal Control

Let us show that the solution of the filtering problem can be obtained as the solution of a certain optimal control problem. Consider the problem of the optimal control for the process

$$
\begin{aligned}
m_u(i) &= \sum_{j=0}^{i-1} u(i,j)y(j+1), \\
m_u(0) &= x(0), \quad i = 1, 2, \ldots, N,
\end{aligned} \tag{5.22}
$$

with the performance functional that has be minimized

$$
\begin{aligned}
J(u) &= \mathbf{E}\left[z_u'(N)Fz_u(N) + \sum_{j=0}^{N-1} z_u'(j)G(j)z_u(j) \right], \\
z_u(j) &= x(j) - m_u(j), \quad j = 0, 1, \ldots, N.
\end{aligned} \tag{5.23}
$$

Here F and $G(j)$ are, respectively, positive definite and positive semidefinite non-random matrices, the stochastic process $y(j)$ is given by the relations (5.2), (5.3) and $u(i,j)$ is a control whose goal is to minimize the performance functional (5.23).

Let $u_0(i,j)$ and $u(i,j)$ be admissible controls and

$$
u_\varepsilon(i,j) = u_0(i,j) + \varepsilon u(i,j), \quad \varepsilon \geq 0. \tag{5.24}
$$

To get the necessary condition for optimality of control, calculate the limit (1.9), (5.24) for the control problem (5.22), (5.23), (5.2). Via (5.23) we have

$$
\begin{aligned}
J_\varepsilon(u_0) &= \frac{1}{\varepsilon}[J(u_\varepsilon) - J(u_0)] \\
&= I_{1\varepsilon} + I_{2\varepsilon},
\end{aligned} \tag{5.25}
$$

where

$$
\begin{aligned}
I_{1\varepsilon} &= \frac{1}{\varepsilon}\mathbf{E}\Big(z_{u_\varepsilon}'(N)Fz_{u_\varepsilon}(N) - z_{u_0}'(N)Fz_{u_0}(N) \Big), \\
I_{2\varepsilon} &= \frac{1}{\varepsilon}\sum_{j=0}^{N-1} \mathbf{E}\Big(z_{u_\varepsilon}'(j)G(j)z_{u_\varepsilon}(j) - z_{u_0}'(j)G(j)z_{u_0}(j) \Big).
\end{aligned}
$$

Via (5.22)–(5.24) for $I_{1\varepsilon}$, we have

$$\begin{aligned}
I_{1\varepsilon} &= \frac{1}{\varepsilon}\mathbf{E}\Big[\big(x(N) - m_{u_\varepsilon}(N)\big)' F\big(x(N) - m_{u_\varepsilon}(N)\big) \\
&\qquad - \big(x(N) - m_{u_0}(N)\big)' F\big(x(N) - m_{u_0}(N)\big)\Big] \\
&= \frac{1}{\varepsilon}\mathbf{E}\Big[- (m_{u_\varepsilon}(N) - m_{u_0}(N))' F(2x(N) - (m_{u_\varepsilon}(N) + m_{u_0}(N)))\Big] \\
&= \frac{1}{\varepsilon}\mathbf{E}\Big[- \varepsilon \sum_{l=0}^{N-1} \Big(u(N,l)y(l+1)\Big)' F\Big(2x(N) - \sum_{k=0}^{N-1}(2u_0(N,k) \\
&\qquad + \varepsilon u(N,k))y(k+1)\Big)\Big] \\
&= \mathbf{E}\Big[- 2\sum_{l=0}^{N-1}\Big(u(N,l)y(l+1)\Big)' Fx(N) \\
&\qquad + 2\sum_{l=0}^{N-1}\Big(u(N,l)y(l+1)\Big)' F\sum_{k=0}^{N-1} u_0(N,k)y(k+1) \\
&\qquad + \varepsilon\sum_{l=0}^{N-1}\Big(u(N,l)y(l+1)\Big)' F\sum_{k=0}^{N-1} u(N,k)y(k+1)\Big].
\end{aligned}$$

So,

$$\begin{aligned}
\lim_{\varepsilon\to 0} I_{1\varepsilon} &= -2\mathbf{E}\Big[\sum_{l=0}^{N-1}\Big(u(N,l)y(l+1)\Big)' Fx(N) \\
&\qquad - \sum_{l=0}^{N-1}\Big(u(N,l)y(l+1)\Big)' F\sum_{k=0}^{N-1} u_0(N,k)y(k+1)\Big] \\
&= -2\Big[\sum_{l=0}^{N-1} Tr\Big(F\mathbf{E}\Big(x(N)y'(l+1)\Big)u'(N,l)\Big) \\
&\qquad - \sum_{l=0}^{N-1} Tr\Big(F\sum_{k=0}^{N-1} u_0(N,k)\mathbf{E}\Big(y(k+1)y'(l+1)\Big)u'(N,l)\Big)\Big] \\
&= -2\Big[\sum_{l=0}^{N-1} Tr\Big(F\Big(\mathbf{E}\Big(x(N)y'(l+1)\Big) \\
&\qquad - \sum_{k=0}^{N-1} u_0(N,k)\mathbf{E}\Big(y(k+1)y'(l+1)\Big)\Big)u'(N,l)\Big)\Big].
\end{aligned} \tag{5.26}$$

Using the similar transformation for $I_{2\varepsilon}$ and noting that via (5.22), (5.23) $z_u(0) = 0$, we obtain

$$\lim_{\varepsilon\to 0} I_{2\varepsilon} = -2\sum_{j=1}^{N-1}\Bigg[\sum_{l=0}^{j-1} Tr\Bigg(G(j)\Bigg(\mathbf{E}\Big(x(j)y'(l+1)\Big) - \sum_{k=0}^{j-1} u_0(j,k)\mathbf{E}\Big(y(k+1)y'(l+1)\Big)\Bigg)u'(j,l)\Bigg)\Bigg]. \tag{5.27}$$

From (5.2), (5.3) similarly to (5.10) it follows that

$$\begin{aligned}\mathbf{E}(x(j)y'(l+1)) &= \mathbf{E}x(j)(A(l)x(l-h)+\xi(l+1))' \\ &= R(j,l-h)A'(l) + Q(j,l+1) \\ &\equiv P(j,l)\end{aligned} \tag{5.28}$$

and via (5.6) similarly to (5.11) we have

$$\begin{aligned}&\sum_{k=0}^{j-1} u_0(j,k)\mathbf{E}(y(k+1)y'(l+1)) \\ &\quad= \sum_{k=0}^{j-1} u_0(j,k)\mathbf{E}(A(k)x(k-h)+\xi(k+1))(A(l)x(l-h)+\xi(l+1))' \\ &\quad= u_0(j,l)S(l+1) + \sum_{k=0}^{j-1} u_0(j,k)\Big(A(k)R(k-h,l-h)A'(l) \\ &\qquad + Q'(l-h,k+1)A'(l) + A(k)Q(k-h,l+1)\Big) \\ &\quad= u_0(j,l)S(l+1) + \sum_{k=0}^{j-1} u_0(j,k)Z(k,l).\end{aligned} \tag{5.29}$$

As a result from (5.25)–(5.29), we obtain

$$J_0(u_0) = -2\Bigg[\sum_{l=0}^{N-1} Tr\Bigg(F\Bigg(P(N,l) - u_0(N,l)S(l+1) - \sum_{k=0}^{N-1} u_0(N,k)Z(k,l)\Bigg)u'(N,l)\Bigg)$$

$$+\sum_{j=1}^{N-1}\left[\sum_{l=0}^{j-1} Tr\left(G(j)\left(P(j,l) - u_0(j,l)S(l+1)\right.\right.\right.$$
$$\left.\left.\left.- \sum_{k=0}^{j-1} u_0(j,k)Z(k,l)\right)u'(j,l)\right)\right] \qquad (5.30)$$

If $u_0(i, j)$ is the optimal control of the problem (5.22), (5.23), (5.2) then $J_0(u_0) \geq 0$. From (5.30) it follows that this inequality holds for arbitrary admissible control $u(j,l)$ if and only if

$$u_0(j,l)S(l+1) + \sum_{k=0}^{j-1} u_0(j,k)Z(k,l) = P(j,l),$$
$$l = 0, 1, \ldots, j-1, \qquad j = 1, 2, \ldots, N, \qquad (5.31)$$

where $P(j,l)$ and $Z(k,l)$ are defined in (5.28) and (5.6), respectively.

In the particular case $F = I$, $G(j) = 0$, we obtain the solution of the problem of optimal (in the mean square sense) estimation of the random variable $x(N)$. By that the estimation error equals

$$\begin{aligned} J(u_0) &= \mathbf{E}\left|x(N) - m_0(N)\right|^2 \\ &= \mathbf{E}Tr\left[x(N)\left(x(N) - m_0(N)\right)'\right] \\ &= Tr\left[R(N,N) - \sum_{j=0}^{N-1} P(N,j)u_0'(N,j)\right], \end{aligned} \qquad (5.32)$$

where $R(N,N)$, $P(N,j)$ and $u_0(N,j)$ are defined in (5.3), (5.28) and (5.31) respectively.

5.1.3 Some Particular Cases

5.1.3.1 In some cases the solution of the equation (5.5) can be obtained in an explicit form. In particular, if $h \geq N$ then via (5.2), (5.3), (5.6), we have $R(N, j-h) = 0$, $Z(i,j) = 0$ for $j < N$. Therefore, from (5.31) for $j = N$ and (5.32) it follows that

$$P(N,j) = Q(N,j+1), \quad u_0(N,j) = Q(N,j+1)S^{-1}(j+1),$$

$$J(u_0) = Tr\left[R(N,N) - \sum_{j=1}^{N} Q(N,j)S^{-1}(j)Q'(N,j)\right].$$

5.1.3.2 Let us suppose now that $h < N$ and the unobservable process $x(i) = \psi(i)x_0$, where $\psi(i)$ is a nonrandom matrix, $\psi(0) = I$, x_0 is a Gaussian random variable independent on $\xi(i)$, $\mathbf{E}x_0 = 0$, $\mathbf{E}x_0x_0' = D_0$. Then from (5.3) it follows that $Q(i, j) = 0$ and the correlation matrix $R(i, j)$ has the form

$$R(i, j) = \psi(i)D_0\psi'(j). \tag{5.33}$$

Via (5.5), (5.6), (5.33) in this case $u_0(j) \equiv 0$ for $0 \le j < h$ and for $h \le j \le N$ it is defined by the equation

$$u_0(j)S(j+1)+\sum_{i=h}^{N-1} u_0(i)A(i)\psi(i-h)D_0\psi'(j-h)A'(j) = \psi(N)D_0\psi'(j-h)A'(j),$$

that can be represented in the form of the second-order Fredholm equation

$$u_0(j) = \left[\psi(N) - \sum_{i=h}^{N-1} u_0(i)A(i)\psi(i-h)\right]D_0\psi'(j-h)A'(j)S^{-1}(j+1). \tag{5.34}$$

Let us define the solution of this equation in the form

$$u_0(j) = \psi(N)F\psi'(j-h)A'(j)S^{-1}(j+1), \quad h \le j \le N, \tag{5.35}$$

where F is a matrix to be found below. Substituting (5.35) into (5.34) and putting

$$\begin{aligned} G &= \sum_{i=h}^{N-1} \psi'(i-h)A_1(i)\psi(i-h) \\ &= \sum_{i=0}^{N-h-1} \psi'(i)A_1(i+h)\psi(i), \\ A_1(i) &= A'(i)S^{-1}(i+1)A(i), \end{aligned} \tag{5.36}$$

we obtain

$$\begin{aligned} &\psi(N)F\psi'(j-h)A'(j)S^{-1}(j+1) \\ &\quad = \psi(N)(I - FG)D_0\psi'(j-h)A'(j)S^{-1}(j+1), \end{aligned}$$

i.e.,

$$F = (I - FG)D_0 \quad \text{or} \quad F = \left(D_0^{-1} + G\right)^{-1}. \tag{5.37}$$

From (5.37) it follows, in particular, that F is a symmetric positive definite matrix.

So, under the above assumptions the optimal estimate is given by formulas (5.35)–(5.37). From (5.6), (5.32), (5.33), (5.35)–(5.37) it follows that the corresponding estimation error equals

$$\begin{aligned}
J(u_0) &= Tr\left[R(N,N) - \sum_{j=h}^{N-1} P(j)u_0'(j)\right] \\
&= Tr\Big[\psi(N)D_0\psi'(N) \\
&\quad - \sum_{j=h}^{N-1} \psi(N)D_0\psi'(j-h)A'(j)\Big(\psi(N)F\psi'(j-h)A'(j)S^{-1}(j+1)\Big)'\Big] \\
&= Tr\left[\psi(N)D_0\psi'(N) - \sum_{j=h}^{N-1} \psi(N)D_0\psi'(j-h)A_1(j)\psi(j-h)F\psi'(N)\right] \\
&= Tr\left[\psi(N)D_0(I-GF)\psi'(N)\right].
\end{aligned}$$

Via (5.37) we have $F' = F = D_0(I - GF)$. So,

$$J(u_0) = Tr\big[\psi(N)F\psi'(N)\big]. \tag{5.38}$$

5.1.4 The Dependence of the Estimation Error on the Magnitude of Delay

Of interest is the question about the dependence of the estimation error on the magnitude of delay in observations. Since $x(i)$ and $\xi(i)$ are independent and $x(j) = 0$ for $j < 0$, the observations (5.2) on the interval $[0, h)$ do not contain any information about the process $x(i)$. Since the total observation time of the process $x(i)$ equals $N - h$ (i.e., i = h, h + 1, ..., N − 1) and decreases as h increases, it seems quite natural, at first glance, to assume that the estimation error, as a function of h, does not decrease as h increases to $N - 1$ and is a constant with respect to h for $h > N - 1$.

Let us consider some scalar examples where one can see that by increase of h the estimation error can both increase and decrease. Considering the estimation error $J(u_0)$ as a function of h, from (5.38), (5.37) we obtain

$$J(h) = \frac{\psi^2(N)}{D_0^{-1} + G(h)}. \tag{5.39}$$

So, the point of the function $G(h)$ maximum is the point of the function $J(h)$ minimum.

Example 5.1 Suppose that A_1 defined in (5.36) is a constant. Then via (5.36)

$$G(h) = \begin{cases} A_1 \sum_{i=0}^{N-h-1} \psi^2(i), & h \le N-1, \\ 0, & h > N-1. \end{cases}$$

It is easy to see that $G(h)$ decreases as the delay h increases to $N-1$. It means that the estimation error $J(h)$ as a function of h increases as the delay h increases to $N-1$ and is a constant if $h > N-1$.

Consider some cases when by an increase of the delay h in the observations the estimation error $J(u_0)$ decreases, that is, the estimation accuracy increases.

Let the unobservable process have the form

$$x(i) = \psi(i)x_0, \quad i \ge 0, \qquad \psi(0) = 1, \tag{5.40}$$

where x_0 is a Gaussian random variable, $\mathbf{E}x_0 = 0$, $\mathbf{E}x_0^2 = \sigma_0^2$. The observable process $y(i)$ is of the type of (5.2), where $\xi(i)$ is a Gaussian stochastic process with independent values which is also independent on x_0, $\mathbf{E}\xi(i) = 0$, $\mathbf{E}\xi^2(i) = \sigma^2(i)$. It is required to estimate the random variable $x(N)$ from the result of observations (5.2).

Via (5.35)–(5.37) the solution of the filtering problem (5.40), (5.2) has the form

$$m(N) = \sum_{j=h}^{N-1} u_0(j)y(j+1),$$

where

$$u_0(j) = \psi(N)F(h)\psi'(j-h)\frac{A(j)}{\sigma^2(j+1)}, \qquad F(h) = \left(\sigma_0^{-2} + G(h)\right)^{-1},$$

$$G(h) = \begin{cases} \sum_{i=0}^{N-h-1} \psi^2(i)A_1(i+h), & h \le N-1, \\ 0, & h > N-1, \end{cases} \qquad A_1(i) = \frac{A^2(i)}{\sigma^2(i+1)}. \tag{5.41}$$

The estimation error as a function of h via (5.39) has the form $J(h) = \psi^2(N)F(h)$. From (5.39) it follows also that the function $J(h)$ decreases if and only if the function $G(h)$ increases, i.e., $J(h+1) < J(h)$ if and only if

$$\begin{aligned} G(h) - G(h+1) &= \sum_{i=0}^{N-h-1} \psi^2(i)A_1(i+h) - \sum_{i=0}^{N-h-2} \psi^2(i)A_1(i+h+1) \\ &= \psi^2(N-h-1)A_1(N-1) \\ &\quad + \sum_{i=0}^{N-h-2} \psi^2(i)\left[A_1(i+h) - A_1(i+h+1)\right] \\ &< 0. \end{aligned} \tag{5.42}$$

Remark 5.1 Note that if the function $A_1(h)$ is a constant then the inequality (5.42) does not hold. It means that $J(h+1) > J(h)$, i.e., $J(h)$ increases if the delay h increases, as it was shown in Example 5.1.

Remark 5.2 Let $\psi(i) \equiv 1$, i.e., $x(i) \equiv x_0$. Then $J(h)$ increases with respect to h for $0 \leq h \leq N-1$ since

$$\begin{aligned} G(h) - G(h+1) &= \sum_{i=0}^{N-h-1} A_1(i+h) - \sum_{i=0}^{N-h-2} A_1(i+h+1) \\ &= A_1(h) > 0. \end{aligned}$$

Example 5.2 Put in the filtering problem (5.40), (5.2)

$$\begin{gathered} \psi(i) = e^{-\alpha i}, \quad A(i) = e^{\beta i}, \quad i \geq 0, \\ \beta > \alpha, \quad \sigma^2(i) = \sigma_0^2 = 1. \end{gathered}$$

Then via (5.41), (5.38) $A_1(i) = A^2(i)\sigma^{-2}(i+1) = e^{2\beta i}$ and

$$J(h) = \frac{e^{-2\alpha N}}{1 + G(h)}, \tag{5.43}$$

where

$$\begin{aligned} G(h) &= \sum_{i=0}^{N-h-1} e^{-2\alpha i} e^{2\beta(i+h)} \\ &= e^{2\beta h} \sum_{i=0}^{N-h-1} e^{2(\beta-\alpha)i} \\ &= e^{2\beta h} \frac{e^{2(N-h)(\beta-\alpha)} - 1}{e^{2(\beta-\alpha)} - 1} \end{aligned} \tag{5.44}$$

if $h \leq N-1$ and $G(h) = 0$ if $h > N-1$.

Considering $G(h)$ as a function of continuous argument, from (5.44) we have

$$\frac{dG(h)}{dh} = \frac{2e^{2\beta h}}{e^{2(\beta-\alpha)} - 1}[\alpha e^{2(N-h)(\beta-\alpha)} - \beta].$$

Therefore, the function $G(h)$ has the maximum in the point

$$h_0 = N - \frac{1}{2(\beta-\alpha)} \ln \frac{\beta}{\alpha}. \tag{5.45}$$

It means that the functional $J(h)$ decreases on the interval $[0, h_0)$ and increases on the interval $(h_0, N-1]$.

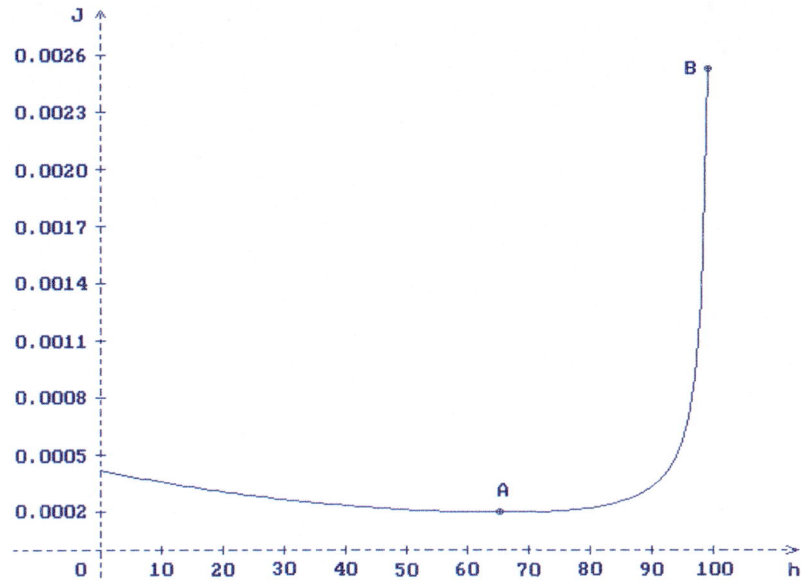

Fig. 5.1 $N = 100$, $\alpha = 0.01$, $\beta = 0.02$, $h \in [0, 99]$

In Fig.5.1 the function $J(h)$ is shown by the following values of the parameters

$$N = 100, \quad \alpha = 0.01, \quad \beta = 0.02, \tag{5.46}$$

for $h \in [0, 99]$. From (5.45), (5.46) it follows that $h_0 = 65.34$. The point A is a point of the minimum of the function $J(h)$, i.e., $J(h_0)$, and the point B is $J(99)$. It is calculated that

$$\begin{aligned}
J(64) &= 0.000200140755,\\
J(65) &= 0.000200009776,\\
J(h_0) &= 0.000200000462,\\
J(66) &= 0.000200035443,\\
J(67) &= 0.000200227558,\\
J(98) &= 0.001316247217,\\
J(99) &= 0.002531650823.
\end{aligned} \tag{5.47}$$

From (5.47) one can see that for integers h the minimal estimation error is $J(65)$.

In Fig.5.2 the same picture is shown for $h \in [0, 115]$. (Here it is supposed that the function $G(h)$ is defined by (5.44) on the interval $h \in [0, N]$ and $G(h) = 0$ for $h \geq N$.) The points A and B as before are $J(h_0)$ and $J(99)$, respectively, and the point C is

$$J(100) = 0.13533528324.$$

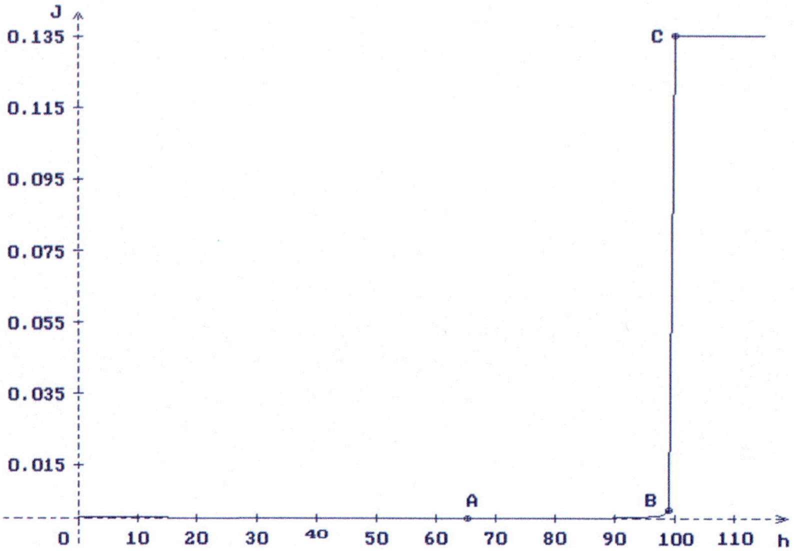

Fig. 5.2 $N = 100$, $\alpha = 0.01$, $\beta = 0.02$, $h \in [0, 115]$

One can see that if the information about the unobservable process is absent absolutely ($h \geq N$) then the estimation error (5.43) increases from $J(99)$ to $J(100)$ very quickly.

Example 5.3 Consider the filtering problem (5.40), (5.2) with

$$\psi(i) = \sqrt{1 - \frac{\alpha i}{N-1}}, \quad A(i) = \sqrt{i}, \quad i = 0, 1, \ldots, N-1,$$
$$\sigma^2(i) = \sigma_0^2 = 1, \quad 0 < \alpha < 1 - \frac{1}{N}.$$

In this case via (5.41) $A_1(i) = i$ and via (5.38), (5.41) we have

$$J(h) = \left(1 - \frac{\alpha N}{N-1}\right)(1 + G(h))^{-1}, \tag{5.48}$$

where

$$\begin{aligned} G(h) &= \sum_{i=0}^{N-h-1} \left(1 - \frac{\alpha i}{N-1}\right)(i+h) \\ &= h + \sum_{i=1}^{N-h-1} \left(1 - \frac{\alpha i}{N-1}\right)(i+h) \end{aligned}$$

$$
\begin{aligned}
&= h + \sum_{i=1}^{N-h-1}(i+h) - \frac{\alpha h}{N-1}\sum_{i=1}^{N-h-1} i - \frac{\alpha}{N-1}\sum_{i=1}^{N-h-1} i^2 \\
&= h + \frac{1}{2}(N+h)(N-h-1) - \frac{\alpha h}{2(N-1)}(N-h)(N-h-1) \\
&\quad - \frac{\alpha}{6(N-1)}(N-h)(N-h-1)(2N-2h-1) \\
&= -\frac{\alpha}{6(N-1)}h^3 - \frac{1}{2}h^2 + \frac{1}{2}\left[1+\alpha\left(N+\frac{1}{3(N-1)}\right)\right]h \\
&\quad + \frac{N}{6}[3(N-1) - \alpha(2N-1)]
\end{aligned}
$$

if $h < N$ and $G(h) = 0$ if $h \geq N$.

Differentiating the function $G(h)$ as a function of continuous argument, we obtain

$$
\frac{dG}{dh} = -\frac{1}{2(N-1)}\left[\alpha h^2 + 2(N-1)h - \left(N - 1 + \alpha\left(N(N-1) + \frac{1}{3}\right)\right)\right].
$$

So, the point

$$
h_0 = \frac{1}{\alpha}\left(\sqrt{(N-1)^2 + \alpha\left[N - 1 + \alpha\left(N(N-1) + \frac{1}{3}\right)\right]} - (N-1)\right) \tag{5.49}
$$

is a point of the function $G(h)$ maximum and at the same time it is the minimum of the function $J(h)$ defined by (5.48). So, the function $J(h)$ decreases if $0 \leq h < h_0$ and increases if $h_0 < h \leq N-1$.

In Fig.5.3 the function $J(h)$ is shown by the following values of the parameters

$$
N = 100, \qquad \alpha = 0.1, \tag{5.50}
$$

for $h \in [0, 100]$. From (5.49), (5.50) it follows that $h_0 = 5.485$. The point A is a point of the minimum of the function $J(h)$, i.e., $J(h_0)$, the points B and C are $J(97)$ and $J(99)$ respectively. It is calculated that

$$
\begin{aligned}
J(4) &= 0.00019402702, \\
J(5) &= 0.00019398558, \\
J(h_0) &= 0.00019398063, \\
J(6) &= 0.00019398621, \\
J(7) &= 0.00019402897, \\
J(97) &= 0.00305051516, \\
J(99) &= 0.00898989899, \\
J(100) &= 0.89898989899.
\end{aligned} \tag{5.51}
$$

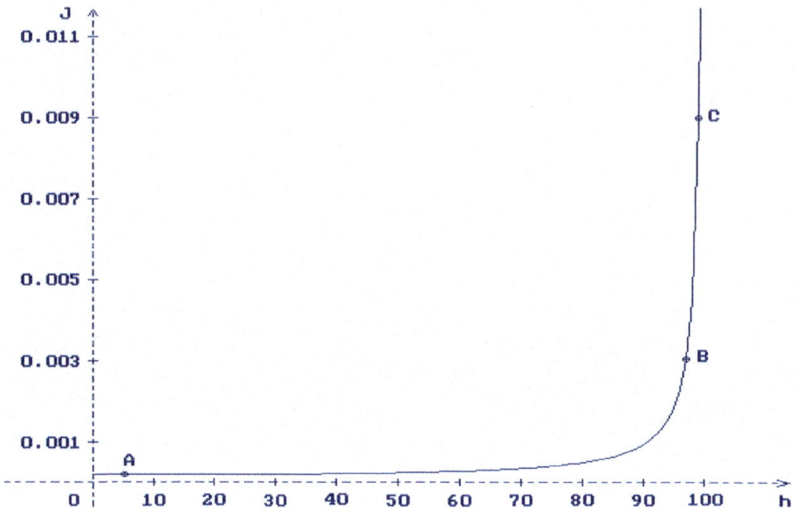

Fig. 5.3 $N = 100, \alpha = 0.1, h \in [0, 100]$

From (5.51) one can see that for integers h the minimal estimation error is $J(5)$. Besides $J(100) = 100J(99)$, i.e., there is enough big jump of the estimation error (5.48) if h increases from $h = 99$ to $h = 100$ because by $h > 99$ the observable process (5.2) is a noise only ($y(i+1) = \xi(i+1)$) and does not depend on the unobservable process $x(i)$.

5.2 Forecasting and Interpolation

In this section it is shown that the solutions of the forecasting and interpolation problems also satisfy the fundamental filtering equation.

5.2.1 Statement and Solution of the Forecasting Problem

The forecasting problem consists in constructing the optimal (in the mean square sense) estimate of the Gaussian variable $x(l)$ at the future moment of time $l > N$ from the observations (5.2) on the segment $0 \le j \le N$. This problem can be reduced to the filtering problem. Really, put

$$A_0(i) = \begin{cases} A(i), & 0 \le i < N, \\ 0, & N \le i \le l, \end{cases} \qquad Q_0(i,j) = \begin{cases} Q(i,j), & 0 \le j \le i < N, \\ 0, & N \le i < l, \end{cases}$$

and consider an auxiliary filtering problem of the random variable $x(l)$ from the results of the observations on the segment $[0, l]$

$$\begin{aligned} y(i+1) &= A_0(i)x(i-h) + \xi(i+1), \\ y(0) &= 0, \quad i = 0, \ldots, l-1, \\ x(j) &= 0, \quad -h \le j < 0, \end{aligned}$$

with $\mathbf{E}x(i)\xi'(j) = Q_0(i, j)$. Since the processes $x(i)$ and $\xi(j)$ are uncorrelated for $i \ge j \ge N$ (and therefore independent because they are Gaussian), the solution of the auxiliary filtering problem is at the same time the solution of the initial forecasting problem. Let $m_0(l)$ be the optimal (in the mean square sense) estimate of the variable $x(l)$ given (similarly to (5.4)) by

$$m_0(l) = \sum_{j=0}^{l-1} u_0(j)y(j+1), \tag{5.52}$$

and $m(l)$ be an estimate of the same form with an arbitrary matrix $u(j)$. Similarly to the proof of Theorem 5.1 we have

$$\mathbf{E}[x(l) - m_0(l)]m'(l) = 0,$$

and using the form (5.52) of the estimates $m_0(l)$ and $m(l)$, we obtain

$$\sum_{j=0}^{l-1}\left(\mathbf{E}x(l)y'(j+1) - \sum_{i=0}^{l-1} u_0(i)\mathbf{E}y(i+1)y'(j+1)\right)u'(j) = 0. \tag{5.53}$$

Put

$$\begin{aligned} P_0(j) &= \begin{cases} R(l, j-h)A'(j) + Q(l, j+1), & 0 \le j < N, \\ 0, & N \le j < l, \end{cases} \\ Z_0(i, j) &= \begin{cases} Z(i, j), & 0 \le j \le i < N, \\ 0, & N \le i < l. \end{cases} \end{aligned} \tag{5.54}$$

Then

$$\begin{aligned} \mathbf{E}x(l)y'(j+1) &= \mathbf{E}x(l)x'(j-h)A_0'(j) + \mathbf{E}x(l)\xi'(j+1) \\ &= R(l, j-h)A_0'(j) + Q_0(l, j+1) \\ &= P_0(j) \end{aligned} \tag{5.55}$$

and

$$\sum_{i=0}^{l-1} u_0(i)\mathbf{E}y(i+1)y'(j+1)$$
$$= \sum_{i=0}^{l-1} u_0(i)\mathbf{E}\Big(A_0(i)x(i-h)+\xi(i+1)\Big)\Big(A_0(j)x(j-h)+\xi(j+1)\Big)'$$
$$= \sum_{i=0}^{l-1} u_0(i)Z_0(i,j) + u_0(j)S(j+1). \tag{5.56}$$

Using the definitions (5.54) of $P_0(j)$ and $Z_0(i, j)$, from (5.53), (5.55), (5.56) we obtain

$$u_0(j)S(j+1) + \sum_{i=0}^{N-1} u_0(i)Z(i,j) = P(j),$$

where $P(j)$ is defined by (5.6) after changing N on l.

So, the solution of the forecasting problem has the form

$$m_0(l) = \sum_{j=0}^{N-1} u_0(j)y(j+1),$$

where $u_0(j)$ is the solution of the equation (5.5), (5.6) after changing in (5.6) N on l.

In particular cases the solution of the forecasting problem can be obtained in the final form. Consider the example from the Sect. 5.1.3 for $h < N$. From (5.34), (5.35) it follows that

$$u_0(j) = \psi(l)F\psi'(j-h)A'(j)S^{-1}(j+1), \tag{5.57}$$

where F is defined by (5.36), (5.37). Similarly to (5.38) the estimation error equals

$$J(u_0) = Tr[\psi(l)F\psi'(l)].$$

5.2.2 Interpolation Problem

The interpolation problem consists in constructing the optimal (in the mean square sense) estimate of the random variable $x(l)$ at a preceding moment of time $l < N$ from the observations (5.2) for $0 \le i \le N$. This problem cannot be directly reduced to the filtering problem but it can be solved similarly to the filtering problem.

In the case of the interpolation problem , the optimal estimate $m_0(l)$ of the random variable $x(l)$ has the representation

$$m_0(l) = \sum_{j=0}^{N-1} u_0(j)y(j+1)$$

too. Since $\mathbf{E}(x(l) - m_0(l))m'(l) = 0$, then similarly to the proof of Theorem 5.1 we obtain that $u_0(j)$ satisfies the condition

$$u_0(j)S(j+1) + \sum_{i=0}^{N-1} u_0(i)Z(i,j) = P(j),$$

where $P(j) = R(l, j-h)A'(j) + Q(l, j+1)$, i.e., the Eqs. (5.5), (5.6) with the replacement in (5.6) N on l.

In particular, if (the example in the Sect. 5.1.3) $h < N$, $Q(i,j) = 0$, $R(i,j)$ is given by (5.33), then $u_0(j)$ is defined by (5.57).

Thus, the solution of all three problems under consideration (the filtering, forecasting and interpolation problems) is given by the Eqs. (5.5), (5.6) with replacement in (5.6) N on l. If $l = N$, then we obtain the solution of the filtering problem, if $l > N$ then we obtain the solution of the forecasting problem, and if $l < N$ then we obtain the solution of the interpolation problem.

5.3 Filtration of Processes Described by Difference Equations

In this section, the filtering problem for a process described by a stochastic difference Volterra equation is considered. To solve this problem two methods are used. An analogue of the Kalman–Bucy filter is constructed, i.e., it is shown that the optimal (in the mean square sense) estimate is defined by a system of four stochastic difference equations. From the other hand, an integral representation of the estimate is also obtained that is defined by the fundamental filtering equation.

5.3.1 An Analogue of the Kalman–Bucy Filter

Consider a partially observable stochastic process $(x(i), y(i))$, defined by the following system of difference equations

$$x(i+1) = x_0 + \sum_{j=0}^{i} a_0(i,j,y_j) + \sum_{j=0}^{i} a_1(i,j,y_j)x(j) + \sum_{j=0}^{i} b(i,j,y_j)\xi(j+1),$$

$$x(0) = x_0, \tag{5.58}$$

$$y(i+1) = A_0(i, y_i) + A_1(i, y_i)x(i) + B(i, y_i)\xi(i+1),$$
$$y(0) = y_0. \tag{5.59}$$

Let $\{\Omega, \mathfrak{F}, \mathbf{P}\}$ be a complete probability space with a family of σ-algebras $\mathfrak{F}_i \subset \mathfrak{F}$, $\mathfrak{F}_i$-measurable standard Gaussian random variables $\xi(i) \in \mathbf{R}^l$, for which $\mathbf{E}\xi(i) = 0$, $\mathbf{E}\xi(i)\xi'(i) = I$ are mutually independent, Gaussian random variables $x_0 \in \mathbf{R}^n$ and $y_0 \in \mathbf{R}^k$ are mutually independent, independent on $\xi(i)$ and such that $\mathbf{E}x_0 = m_0$, $\mathbf{E}(x_0 - m_0)(x_0 - m_0)' = \gamma_0$, $\mathbf{E}y_0 = 0$.

It is supposed that for an arbitrary $\mathfrak{F}_j^y$-measurable function $\varphi(j) \in \mathbf{R}^k$ the functional

$$a_0(i, j, \varphi) = \left(a_1^0(i, j, \varphi), \ldots, a_n^0(i, j, \varphi)\right),$$

the matrices

$$a_1(i, j, \varphi) = \left\|a_{pq}^1(i, j, \varphi)\right\|$$

and

$$b(i, j, \varphi) = \left\|b_{pq}(i, j, \varphi)\right\|$$

of the dimensions $n \times n$ and $n \times l$, respectively, depend on values of the function $\varphi(s)$ for $s = 0, 1, \ldots, j$, the functional

$$A_0(i, \varphi) = \left(A_1^0(i, \varphi), \ldots, A_m^0(i, \varphi)\right),$$

the matrices

$$A_1(i, \varphi) = \left\|A_{pq}^1(i, \varphi)\right\|$$

and

$$B(i, \varphi) = \left\|B_{pq}(i, \varphi)\right\|$$

of the dimensions $k \times n$ and $k \times l$, respectively, depend on the values of the function $\varphi(j)$ for $j = 0, 1, \ldots, i$ and satisfy the conditions that provide [130] conditional gaussianity of the processes $x(i)$ and $y(i)$: by the probability 1

$$\left|a_{pq}^1(i, j, \varphi)\right| \leq c, \qquad \left|A_{pq}^1(i, \varphi)\right| \leq c,$$

and

$$\mathbf{E}\left|a_p^0(i, j, \varphi)\right|^2 < \infty, \quad \mathbf{E}\left|b_{pq}(i, j, \varphi)\right|^2 < \infty, \quad i, j \in Z,$$
$$\mathbf{E}\left|A_p^0(i, \varphi)\right|^2 < \infty, \quad \mathbf{E}\left|B_{pq}(i, \varphi)\right|^2 < \infty, \quad i \in Z.$$

Let us introduce the following notation:

$$x^i(j+1) = x_0 + \sum_{l=0}^{j} a_0(i,l,y_l) + \sum_{l=0}^{j} a_1(i,l,y_l)x(l) + \sum_{l=0}^{j} b(i,l,y_l)\xi(l+1),$$
$$j = 0, 1, \dots, i-1, \quad x^i(0) = x_0. \tag{5.60}$$

Note that for $j = i$ (5.60) coincides with (5.58) and rewrite (5.60) in the form

$$x^i(j+1) = x^i(j) + a_0(i,j,y_j) + a_1(i,j,y_j)x(j) + b(i,j,y_j)\xi(j+1). \tag{5.61}$$

Put also

$$\begin{aligned} m(i) &= \mathbf{E}_i^y x(i), \quad m^i(j) = \mathbf{E}_j^y x^i(j), \\ \delta(i) &= x(i) - m(i), \quad \delta^i(j) = x^i(j) - m^i(j), \\ K(i,j) &= \mathbf{E}_i^y \delta(i)\delta'(j), \quad f(i,j) = \mathbf{E}_j^y \delta^i(j)\delta'(j), \quad i \geq j, \quad \gamma(i) = K(i,i), \end{aligned} \tag{5.62}$$

and

$$\begin{aligned} D(i,j,y_j) &= \left[(f(i,j) + a_1(i,j,y_j)\gamma(j))A_1'(j,y_j) + b(i,j,y_j)B'(j,y_j)\right] \\ &\quad \times \left[A_1(j,y_j)\gamma(j)A_1'(j,y_j) + B(j,y_j)B'(j,y_j)\right]^{\oplus}, \\ P(i,j,y_j) &= a_1(i,j,y_j) - D(i,j,y_j)A_1(j,y_j), \\ Q(i,j,y_j) &= b(i,j,y_j) - D(i,j,y_j)B(j,y_j). \end{aligned} \tag{5.63}$$

Theorem 5.2 *Let a partially observable process (x(i), y(i)) is defined by the Eqs.* (5.58), (5.59) *and the above assumptions hold. Then* $m(i)$, $f(i,j)$, $\gamma(i)$ *and* $K(i,j)$ *are defined by the equations*

$$\begin{aligned} m(i+1) &= m_0 + \sum_{j=0}^{i} a_0(i,j,y_j) + \sum_{j=0}^{i} a_1(i,j,y_j)m(j) \\ &\quad + \sum_{j=0}^{i} D(i,j,y_j)\Big[y(j+1) - A_0(j,y_j) - A_1(j,y_j)m(j)\Big], \end{aligned} \tag{5.64}$$

$$\begin{aligned} f(i,j+1) = \gamma_0 &+ \sum_{l=0}^{j} \Big[P(i,l,y_l)f'(j,l) + f(i,l)P'(j,l,y_l) \\ &+ P(i,l,y_l)\gamma(l)P'(j,l,y_l)\Big] + \sum_{l=0}^{j} Q(i,l,y_l)Q'(j,l,y_l), \end{aligned} \tag{5.65}$$

$$\gamma(i+1) = \gamma_0 + \sum_{l=0}^{i} \Big[P(i,l,y_l) f'(i,l) + f(i,l) P'(i,l,y_l) + P(i,l,y_l)\gamma(l) P'(i,l,y_l) \Big] + \sum_{l=0}^{i} Q(i,l,y_l) Q'(i,l,y_l), \tag{5.66}$$

$$K(i+1,j) = f(i,j+1) + \sum_{l=j+1}^{i} P(i,l,y_l) K(l,j). \tag{5.67}$$

Proof Let us obtain the equation for $m(i)$. From (5.61), (5.59) via (5.62), we have

$$\mathbf{E}_j^y x^i(j+1) = m^i(j) + a_0(i,j,y_j) + a_1(i,j,y_j) m(j), \tag{5.68}$$

$$\mathbf{E}_j^y y(j+1) = A_0(j,y_j) + A_1(j,y_j) m(j), \tag{5.69}$$

Subtracting (5.68) from (5.61) and (5.69) from (5.59) and using (5.62), we obtain

$$x^i(j+1) - \mathbf{E}_j^y x^i(j+1) = \delta^i(j) + a_1(i,j,y_j)\delta(j) + b(i,j,y_j)\xi(j+1), \tag{5.70}$$

$$y(j+1) - \mathbf{E}_j^y y(j+1) = A_1(j,y_j)\delta(j) + B(j,y_j)\xi(j+1). \tag{5.71}$$

To use the theorem about normal correlation (Theorem 1.1) let us calculate

$$D_{12} = \mathbf{E}_j^y \Big[\big(x^i(j+1) - \mathbf{E}_j^y x^i(j+1) \big) \big(y(j+1) - \mathbf{E}_j^y y(j+1) \big)' \Big], \tag{5.72}$$

and

$$D_{22} = \mathbf{E}_j^y \Big[\big(y(j+1) - \mathbf{E}_j^y y(j+1) \big) \big(y(j+1) - \mathbf{E}_j^y y(j+1) \big)' \Big]. \tag{5.73}$$

From (5.70)–(5.73), (5.62), we obtain

$$\begin{aligned} D_{12} &= \mathbf{E}_j^y \Big[\big(\delta^i(j) + a_1(i,j,y_j)\delta(j) + b(i,j,y_j)\xi(j+1) \big) \\ &\quad \times \big(A_1(j,y_j)\delta(j) + B(j,y_j)\xi(j+1) \big)' \Big] \\ &= f(i,j) A_1'(j,y_j) + a_1(i,j,y_j)\gamma(j) A_1'(j,y_j) + b(i,j,y_j) B'(j,y_j), \end{aligned} \tag{5.74}$$

and

$$\begin{aligned} D_{22} &= \mathbf{E}_j^y \Big[\big(A_1(j,y_j)\delta(j) + B(j,y_j)\xi(j+1) \big) \\ &\quad \times \big(A_1(j,y_j)\delta(j) + B(j,y_j)\xi(j+1) \big)' \Big] \\ &= A_1(j,y_j)\gamma(j) A_1'(j,y_j) + B(j,y_j) B'(j,y_j). \end{aligned} \tag{5.75}$$

From (5.63), (5.74), (5.75) it follows that $D(i, j, y_j) = D_{12}D_{22}^{\oplus}$. So, via Theorem 1.1 from (5.68), (5.69) we have

$$\begin{aligned} m^i(j+1) &= m^i(j) + a_0(i, j, y_j) + a_1(i, j, y_j)m(j) \\ &\quad + D(i, j, y_j)\big[y(j+1) - A_0(j, y_j) - A_1(j, y_j)m(j)\big]. \end{aligned} \tag{5.76}$$

Summing (5.76) over $j = 0, \ldots, k$, we obtain

$$\begin{aligned} \sum_{j=0}^{k} m^i(j+1) &= \sum_{j=0}^{k} m^i(j) + \sum_{j=0}^{k} a_0(i, j, y_j) + \sum_{j=0}^{k} a_1(i, j, y_j)m(j) \\ &\quad + \sum_{j=0}^{k} D(i, j, y_j)\big[y(j+1) - A_0(j, y_j) - A_1(j, y_j)m(j)\big]. \end{aligned} \tag{5.77}$$

Note that

$$\sum_{j=0}^{k} m^i(j+1) - \sum_{j=0}^{k} m^i(j) = m^i(k+1) - m_0 \tag{5.78}$$

and for $k = i$ via (5.62), (5.60), (5.58) we have

$$m^i(i+1) = \mathbf{E}_{i+1}^{y} x^i(i+1) = \mathbf{E}_{i+1}^{y} x(i+1) = m(i+1). \tag{5.79}$$

The Eq. (5.64) follows from (5.77)–(5.79).

Let us get now the Eqs. (5.65), (5.66) for $\gamma(i)$ and $f(i, j)$. Using (5.69), (5.71), rewrite (5.76) in the form

$$\begin{aligned} m^i(j+1) &= m^i(j) + a_0(i, j, y_j) + a_1(i, j, y_j)m(j) \\ &\quad + D(i, j, y_j)\big[A_1(j, y_j)\delta(j) + B(j, y_j)\xi(j+1)\big]. \end{aligned} \tag{5.80}$$

Subtracting (5.80) from (5.61), via (5.62), (5.63), we obtain

$$\delta^i(j+1) = \delta^i(j) + P(i, j, y_j)\delta(j) + Q(i, j, y_j)\xi(j+1). \tag{5.81}$$

Then

$$\begin{aligned} \delta^i(l+1)\big(\delta^j(l+1)\big)' &= (\delta^i(l) + P(i, l, y_l)\delta(l) + Q(i, l, y_l)\xi(l+1)) \\ &\quad \times (\delta^j(l) + P(j, l, y_l)\delta(l) + Q(j, l, y_l)\xi(l+1))' \\ &= \delta^i(l)(\delta^j(l))' + P(i, l, y_l)\delta(l)(\delta^j(l))' \\ &\quad + \delta^i(l)\delta'(l)P'(j, l, y_l) + P(i, l, y_l)\delta(l)\delta'(l)P'(j, l, y_l) \end{aligned}$$

$$
\begin{aligned}
&+ Q(i,l,y_l)\xi(l+1)(\delta^j(l) + P(j,l,y_l)\delta(l))' \\
&+ (\delta^i(l) + P(i,l,y_l)\delta(l))(Q(j,l,y_l)\xi(l+1))' \\
&+ Q(i,l,y_l)\xi(l+1)\xi'(l+1)Q'(j,l,y_l)
\end{aligned}
$$

and after calculating the expectation

$$
\begin{aligned}
\mathbf{E}_l^y \delta^i(l+1)(\delta^j(l+1))' = \mathbf{E}_l^y \delta^i(l)(\delta^j(l))' &+ P(i,l,y_l)\mathbf{E}_l^y \delta(l)(\delta^j(l))' + \mathbf{E}_l^y \delta^i(l)\delta'(l)P'(j,l,y_l) \\
&+ P(i,l,y_l)\mathbf{E}_l^y \delta(l)\delta'(l)P'(j,l,y_l) \\
&+ Q(i,l,y_l)Q'(j,l,y_l).
\end{aligned}
$$

From this via (5.62) we have

$$
\begin{aligned}
\mathbf{E}_l^y \delta^i(l+1)(\delta^j(l+1))' = \mathbf{E}_l^y \delta^i(l)(\delta^j(l))' &+ P(i,l,y_l)f'(j,l) + f(i,l)P'(j,l,y_l) \\
&+ P(i,l,y_l)\gamma(l)P'(j,l,y_l) + Q(i,l,y_l)Q'(j,l,y_l)
\end{aligned}
$$

and via Theorem 1.1 at the same time

$$
\begin{aligned}
\mathbf{E}_{l+1}^y \delta^i(l+1)(\delta^j(l+1))' = \mathbf{E}_l^y \delta^i(l)(\delta^j(l))' &+ P(i,l,y_l)f'(j,l) + f(i,l)P'(j,l,y_l) \\
&+ P(i,l,y_l)\gamma(l)P'(j,l,y_l) + Q(i,l,y_l)Q'(j,l,y_l).
\end{aligned}
$$

Summing the obtained equality over $l = 0, \ldots, k$, we obtain

$$
\begin{aligned}
\mathbf{E}_{k+1}^y \delta^i(k+1)(\delta^j(k+1))' = \mathbf{E}_0^y \delta^i(0)(\delta^j(0))' &+ \sum_{l=0}^{k}[P(i,l,y_l)f'(j,l) + f(i,l)P'(j,l,y_l) \\
&+ P(i,l,y_l)\gamma(l)P'(j,l,y_l) + Q(i,l,y_l)Q'(j,l,y_l)].
\end{aligned} \tag{5.82}
$$

Note that via (5.58), (5.60), (5.62) we have

$$
\mathbf{E}_0^y \delta^i(0)(\delta^j(0))' = \mathbf{E}_0^y \delta(0)\delta'(0) = \gamma_0
$$

and besides for $k = j$ in (5.82) we obtain

$$
\begin{aligned}
\mathbf{E}_{j+1}^y \delta^i(j+1)\big(\delta^j(j+1)\big)' &= \mathbf{E}_{j+1}^y \delta^i(j+1)\delta'(j+1) \\
&= f(i,j+1).
\end{aligned}
$$

So, from this and (5.82) we get (5.65).

Note also that via (5.62), (5.79)

$$\begin{aligned} f(i,i+1) &= \mathbf{E}^y_{i+1}\delta^i(i+1)\delta'(i+1) \\ &= \mathbf{E}^y_{i+1}\delta(i+1)\delta'(i+1) \\ &= K(i+1,i+1) \\ &= \gamma(i+1). \end{aligned}$$

So, putting in (5.65) $j=i$, we obtain (5.66).

To get the equation for $K(i,j)$ put

$$\chi^j(l) = \begin{cases} 1 & \text{if } l \le j, \\ 0 & \text{if } l > j, \end{cases} \qquad \sigma^j(l) = \begin{cases} \delta^j(l) & \text{if } l \le j, \\ \delta(j) & \text{if } l > j, \end{cases} \tag{5.83}$$

and consider the process

$$\sigma^j(l+1) = \sigma^j(l) + \chi^j(l)\Big(P(j,l,y_l)\delta(l) + \chi^j(l)Q(j,l,y_l)\Big). \tag{5.84}$$

Using (5.81), (5.84), similarly to (5.82) we obtain

$$\begin{aligned} \mathbf{E}^y_{k+1}\delta^i(k+1)(\sigma^j(k+1))' &= \mathbf{E}^y_0\delta^i(0)(\sigma^j(0))' \\ &+ \sum_{l=0}^{k}\Big[P(i,l,y_l)\mathbf{E}^y_l\delta(l)(\sigma^j(l))' + \Big(f(i,l)P'(j,l,y_l) \\ &+ P(i,l,y_l)\gamma(l)P'(j,l,y_l) \\ &+ Q(i,l,y_l)Q'(j,l,y_l)\Big)(\chi^j(l))'\Big]. \end{aligned} \tag{5.85}$$

Note that via (5.58), (5.60), (5.62), (5.83), we have

$$\mathbf{E}^y_0\delta^i(0)(\sigma^j(0))' = \mathbf{E}^y_0\delta(0)\delta'(0) = \gamma_0,$$

and besides for $k=i\ge j$ in (5.85) we obtain

$$\begin{aligned} \mathbf{E}^y_{i+1}\delta^i(i+1)\big(\sigma^j(i+1)\big)' &= \mathbf{E}^y_{i+1}\delta(i+1)\delta'(j) \\ &= K(i+1,j). \end{aligned}$$

If $l \le j$ then

$$\mathbf{E}^y_l\delta(l)(\sigma^j(l))' = \mathbf{E}^y_l\delta(l)(\delta^j(l))' = f'(j,l),$$

if $l > j$ then

$$\mathbf{E}^y_l\delta(l)(\sigma^j(l))' = \mathbf{E}^y_l\delta(l)\delta'(j) = K(l,j).$$

So, from (5.85) for $k = i$ it follows that

$$
\begin{aligned}
K(i+1,j) = \gamma_0 &+ \sum_{l=0}^{j} \big[P(i,l,y_l) f'(j,l) + f(i,l) P'(j,l,y_l) \\
&+ P(i,l,y_l)\gamma(l) P'(j,l,y_l) + Q(i,l,y_l) Q'(j,l,y_l)\big] \\
&+ \sum_{l=j+1}^{i} P(i,l,y_l) K(l,j).
\end{aligned}
$$

From this and (5.65) we obtain (5.67). The proof is completed.

5.3.2 An Integral Representation of the Estimate

Consider the problem of constructing of the optimal (in the mean square sense) estimate $m_0(N)$, using the observations (5.2) of the variable $x(N)$ given by the Volterra equation

$$
x(i+1) = \xi_0(i+1) + \sum_{j=0}^{i} a(i,j) x(j), \quad 0 \le i \le N-1, \quad x(0) = \xi_0(0). \tag{5.86}
$$

Here $\xi_0(i)$ is an $\mathfrak{F}_i$-measurable Gaussian process such that $\mathbf{E}\xi_0(i) = 0$, $\mathbf{E}\xi_0(i)\xi_0'(j) = N_0(i,j)$, $\mathbf{E}\xi_0(i)\xi'(j) = N_1(i,j)$ and $N_1(i,j) = 0$ for $j \ge i$.

Let us obtain the fundamental filtering equation for the considered optimal filtering problem (5.86), (5.2). Using the resolvent $R_a(i,j)$ of the kernel $a(i,j)$, represent the solution of the equation (5.86) in the form

$$
x(i+1) = \xi_0(i+1) + \sum_{k=0}^{i} R_a(i,k)\xi_0(k). \tag{5.87}
$$

For arbitrary function $f(i)$ put

$$
\psi_a(i, f(\cdot)) = \begin{cases} f(i) + \sum_{k=0}^{i-1} R_a(i-1,k) f(k), & i \ge 1, \\ f(0), & i = 0, \\ 0, & i < 0. \end{cases} \tag{5.88}
$$

Then via (5.86)–(5.88) the solution (5.87) of the Eq. (5.86) can be represented in the form

$$
x(i) = \psi_a(i, \xi_0(\cdot)), \qquad i \ge 0. \tag{5.89}
$$

From (5.3), (5.88), (5.89) it follows that

$$
\begin{aligned}
R(i,j) &= \mathbf{E}x(i)x'(j) \\
&= \mathbf{E}\psi_a(i,\xi_0(\cdot))\psi'_a(j,\xi_0(\cdot)) \\
&= \mathbf{E}\left(\xi_0(i) + \sum_{k=0}^{i-1} R_a(i-1,k)\xi_0(k)\right)\left(\xi_0(j) + \sum_{l=0}^{j-1} R_a(j-1,l)\xi_0(l)\right)' \\
&= N_0(i,j) + \sum_{l=0}^{j-1} N_0(i,l)R'_a(j-1,l) \\
&\quad + \sum_{k=0}^{i-1} R_a(i-1,k)\left(N_0(k,j) + \sum_{l=0}^{j-1} N_0(k,l)R'_a(j-1,l)\right) \\
&= \left(N_0(j,i) + \sum_{l=0}^{j-1} R_a(j-1,l)N_0(l,i)\right)' \\
&\quad + \sum_{k=0}^{i-1} R_a(i-1,k)\left(N_0(j,k) + \sum_{l=0}^{j-1} R_a(j-1,l)N_0(l,k)\right)'.
\end{aligned}
$$

As a result from this via (5.88), we obtain

$$
\begin{aligned}
R(i,j) &= \psi'_a(j,N_0(\cdot,i)) + \sum_{k=0}^{i-1} R_a(i-1,k)\psi'_a(j,N_0(\cdot,k)) \\
&= \psi_a(i,\psi'_a(j,N_0(\cdot,\cdot))).
\end{aligned}
\tag{5.90}
$$

Similarly, via (5.3), (5.88) we have

$$
\begin{aligned}
Q(i,j) &= \mathbf{E}x(i)\xi'(j) \\
&= \mathbf{E}\psi_a(i,\xi_0(\cdot))\xi'(j) \\
&= \psi_a(i,N_1(\cdot,j)).
\end{aligned}
\tag{5.91}
$$

From (5.4)–(5.6), (5.90), (5.91), we obtain the following statement.

Corollary 5.1 *The optimal (in the mean square sense) estimate $m_0(N)$ of the filtering problem* (5.86), (5.2) *is defined by the representation*

$$
m_0(N) = \sum_{j=0}^{N-1} u_0(j)y(j+1),
$$

where $u_0(j)$ is a unique solution of the equation

$$u_0(j)S(j+1)+\sum_{i=0}^{N-1}u_0(i)Z(i,j)=P(j)$$

with

$$\begin{aligned}
P(j)&=\psi_a(N,\psi_a'(j-h,0,N_0(\cdot,\cdot)))A'(j)\\
&\quad+\psi_a(N,N_1(\cdot,j+1)),\\
Z(i,j)&=A(i)\psi_a(i-h,\psi_a'(j-h,N_0(\cdot,\cdot)))A'(j)\\
&\quad+A(i)\psi_a'(i-h,N_1(\cdot,j+1))\\
&\quad+\psi_a'(j-h,N_1(\cdot,i+1))A'(j).
\end{aligned}$$

5.4 Importance of Research of Difference Volterra Equations

In order to stress an importance of research of difference Volterra equations let us show that the solution of the optimal filtering problem for stochastic difference Volterra equation cannot be obtained as a difference approximation of the solution of a similar problem for stochastic integral equation.

More exactly, let us consider an example, in which the solution of the optimal filtering problem for the equation with discrete time does not coincide with the difference analogue of solution of the similar optimal filtering problem for equation with continuous time.

Consider, for instance, the simple optimal filtering problem for stochastic Volterra integral equation. Let $(x(t),y(t))$ be a partially observable process defined by the following system of integral equations

$$\begin{aligned}
x(t)&=x_0+\int_0^t a(t,s)x(s)ds+\int_0^t b(t,s)dw(s),\\
y(t)&=y_0+\int_0^t A(s)x(s)ds+\int_0^t B(s)dw(s).
\end{aligned}\tag{5.92}$$

It is assumed here that a complete probability space $\{\Omega,\mathfrak{F},\mathbf{P}\}$ with a family of σ-algebras $\mathfrak{F}_t\in\mathfrak{F}$, an $\mathfrak{F}_t$-measurable scalar standard Wiener process $w(t)$, Gaussian random variables x_0 and y_0 such that $\mathbf{E}x_0=0$, $\mathbf{E}x_0^2=\gamma_0$, $\mathbf{E}y_0=0$, mutually independent and independent on $w(t)$, are given. The symbol $\mathfrak{F}_t^y$ denotes the σ-algebra generated by values of the process $y(s)$, $s\le t$, $\mathbf{E}_t^y=\mathbf{E}\{\cdot/\mathfrak{F}_t^y\}$. The matrices $a(t,s)$, $b(t,s)$, $A(s)$, $B(s)$, and $(B(s)B'(s))^{-1}$ have appropriate dimensions and are uniformly bounded.

The analogue of the Kalman–Busy filter for this problem is defined [9, 109] by the system of four stochastic integral equations, where the first one that defines $m(t) = \mathbf{E}_t^y x(t)$ has the form:

$$m(t) = m(0) + \int_0^t a(t,s)m(s)\mathrm{d}s + \int_0^t D(t,s)B_0(s)\,[\mathrm{d}y(s) - A(s)m(s)\mathrm{d}s] \tag{5.93}$$

with

$$D(t,s) = f(t,s)A'(s) + b(t,s)B'(s). \tag{5.94}$$

To construct the different analogues of the Eqs. (5.92) and (5.93) put

$$t = \Delta i, \qquad s = \Delta j, \qquad \bar{y}(i+1) = y(i+1) - y(i),$$
$$\xi(i+1) = \frac{1}{\sqrt{\Delta}}(w((i+1)\Delta) - w(i\Delta)),$$

where $\Delta > 0$ is the step of discretization. Note that $\mathbf{E}\xi(i+1) = 0$, $\mathbf{E}\xi^2(i+1) = 1$.

The difference analogues of the Eqs. (5.92), (5.93), respectively, have the forms

$$x(i+1) = x_0 + \sum_{j=0}^{i} a(i,j)\Delta x(j) + \sum_{j=0}^{i} b(i,j)\sqrt{\Delta}\xi(j+1),$$
$$\bar{y}(i+1) = A(i)\Delta x(i) + B(i)\sqrt{\Delta}\xi(i+1) \tag{5.95}$$

and

$$m(i+1) = m(0) + \sum_{j=0}^{i} a(i,j)\Delta m(j) + \sum_{j=0}^{i} D(i,j)B_0(j)\,[\bar{y}(j+1) - \Delta A(j)m(j)], \tag{5.96}$$

where

$$B_0(j) = \frac{1}{\Delta}[B(j)B'(j)]^{-1},$$
$$D(i,j) = \Delta[f(i,j)A'(j) + b(i,j)B'(j)],$$

and

$$D(i,j)B_0(j) = [f(i,j)A'(j) + b(i,j)B'(j)][B(j)B'(j)]^{-1}. \tag{5.97}$$

On the other hand from (5.64), (5.63) it follows that the equation for the optimal estimate $m(i)$ of the solution of the problem (5.95) has the form

$$\begin{aligned} m(i+1) = m_0 + \sum_{j=0}^{i} a(i,j)\Delta m(j) \\ + \sum_{j=0}^{i} \bar{D}(i,j)\bar{B}_0(j)\,[\bar{y}(j+1) - \Delta A(j)m(j)], \end{aligned} \tag{5.98}$$

where

$$\bar{B}_0(j) = \frac{1}{\Delta}\big[B(j)B'(j) + A(j)\gamma(j)A'(j)\Delta\big]^{-1},$$
$$\bar{D}(i,j) = \Delta\big[f(i,j)A'(j) + b(i,j)B'(j) + a(i,j)\gamma(j)A'(j)\Delta\big],$$

and

$$\begin{aligned} \bar{D}(i,j)\bar{B}_0(j) = \big[f(i,j)A'(j) + b(i,j)B'(j) + a(i,j)\gamma(j)A'(j)\Delta\big] \\ \times [B(j)B'(j) + A(j)\gamma(j)A'(j)\Delta]^{-1}. \end{aligned} \tag{5.99}$$

Comparing (5.97) with (5.99), it is easy to see that the Eqs. (5.98), (5.99) has a significant difference from the Eqs. (5.96), (5.97). By that each from the Eqs. (5.96) and (5.98) converges to the Eq. (5.93) if $\Delta \to 0$.

Chapter 6
Optimal Control of Stochastic Difference Volterra Equations by Incomplete Information

In this chapter, two methods for solution of the optimal control problem of a partially observable linear stochastic process with a quadratic performance functional are considered.

By the first method that is the separation method the considered optimal control problem is separated into two problems: the optimal filtering problem and the optimal control problem for some auxiliary system [130, 205]. The auxiliary system is described by stochastic difference Volterra equation and the optimal control problem for this system is solved by the method considered in Chap. 2. For the solution of the problem of optimal filtering, the method given in Chap. 5 is used.

Another way of the solution of the optimal control problem by incomplete information in the case when an unobservable process has a delay is an analogue of the method of integral representations that was firstly proposed for stochastic integral equations [9, 109].

For a quasilinear stochastic difference Volterra equation, the zeroth approximation to the optimal control is constructed.

6.1 Separation Method

Let $\{\Omega, \mathfrak{F}, P\}$ be a basic probability space, $\mathfrak{F}_i \subset \mathfrak{F}$ be a family of σ-algebras, $Z = \{0, 1, \dots, N\}$, H be a space of $\mathfrak{F}_i$-adapted random variables $x(i) \in \mathbf{R}^n$, $i \in Z$, with the norm

$$\|x\|_i^2 = \max_{0 \le j \le i} \mathbf{E}|x(j)|^2 < \infty.$$

Consider the optimal control problem for the stochastic difference Volterra equation

$$x(i+1) = x_0 + \sum_{j=0}^{i} a_0(i,j)u(j) + \sum_{j=0}^{i} a_1(i,j)x(j) + a_2(i)\eta + \sum_{j=0}^{i} b(i,j)\xi(j+1),$$
$$x(0) = x_0, \quad i = 0, 1, \dots, N-1, \tag{6.1}$$

L. Shaikhet, *Optimal Control of Stochastic Difference Volterra Equations*, Studies in Systems, Decision and Control 17, DOI 10.1007/978-3-319-13239-6_6

with the observable process

$$y(i+1) = A(i)x(i) + B(i)\xi(i+1), \qquad y(0) = 0, \tag{6.2}$$

and the performance functional

$$J(u) = \mathbf{E}\left[x'(N)Fx(N) + \sum_{j=0}^{N-1} u'(j)G(j)u(j)\right]. \tag{6.3}$$

Here $x(i) \in \mathbf{R}^n$, $y(i) \in \mathbf{R}^k$, $\xi(i) \in \mathbf{R}^l$ are $\mathfrak{F}_i$-measurable mutually independent Gaussian variables such that $\mathbf{E}\xi(i) = 0$, $\mathbf{E}\xi(i)\xi'(i) = I$, $u(j) \in \mathbf{R}^m$, $a_0(i,j)$, $a_1(i,j), a_2(i), b(i,j), A(i)$, and $B(j)$ are nonrandom matrices of dimensions $n \times m$, $n \times n, n \times l_1, n \times l, k \times n$ and $k \times l$, respectively. The vector x_0 (the initial value of the process $x(i)$) and the unknown parameter $\eta \in \mathbf{R}^{l_1}$ are mutually independent and independent on $\xi(i)$ Gaussian variables with expectations $m_1(0) = \mathbf{E}x_0$, $m_2(0) = \mathbf{E}\eta$ and covariation matrices $\gamma_0 = \mathbf{E}(x_0 - m_1(0))(x_0 - m_1(0))'$ and $\gamma_1 = \mathbf{E}(\eta - m_2(0))(\eta - m_2(0))'$, respectively, $a_2(0) = 0$, F is a positive semidefinite matrix, $G(j)$ is a positive definite matrix of the dimensions $n \times n$ and $m \times m$, respectively.

6.1.1 Solution of the Filtering Problem

Represent the system (6.1), (6.2) in the form

$$\bar{x}(i+1) = \bar{x}_0 + \sum_{j=0}^{i} \bar{a}_0(i,j)u(j) + \sum_{j=0}^{i} \bar{a}_1(i,j)\bar{x}(j) + \sum_{j=0}^{i} \bar{b}(i,j)\xi(j+1), \tag{6.4}$$

$$y(i+1) = \bar{A}(i)\bar{x}(i) + B(i)\xi(i+1), \qquad y(0) = 0. \tag{6.5}$$

Here

$$\begin{gathered}
\bar{x}(i) = \begin{pmatrix} x(i) \\ \eta(i) \end{pmatrix}, \quad \bar{x}_0 = \begin{pmatrix} x_0 \\ \eta \end{pmatrix}, \quad \bar{a}_0(i,j) = \begin{pmatrix} a_0(i,j) \\ 0 \end{pmatrix}, \\
\bar{a}_1(i,j) = \begin{pmatrix} a_1(i,j) & \tilde{a}_2(j) \\ 0 & 0 \end{pmatrix}, \quad \bar{b}(i,j) = \begin{pmatrix} b(i,j) \\ 0 \end{pmatrix}, \\
\tilde{a}_2(j) = a_2(j) - a_2(j-1), \quad j = 1, \ldots, i, \quad \tilde{a}_2(0) = a_2(0), \\
\eta(i) = \eta, \quad \bar{A}(i) = \big(A(i), 0\big).
\end{gathered} \tag{6.6}$$

In compliance with the separation principle [205], the control problem (6.1)–(6.3) has been solved in two steps: first the filtering problem (6.4), (6.5) for a fixed control $u(j)$ and then some auxiliary control problem. Let us describe these two steps.

Let $\mathfrak{F}_i^y$ be σ-algebra generated by the process $y(j)$, $0 \le j \le i$, $\mathbf{E}_i^y = \mathbf{E}\big(\cdot/\mathfrak{F}_i^y\big)$. Put

$$x^i(j+1) = x_0 + \sum_{l=0}^{j} a_0(i,l)u(l) + \sum_{l=0}^{j} a_1(i,l)x(l) + a_2(j)\eta + \sum_{l=0}^{j} b(i,l)\xi(l+1),$$
$$0 \le j \le i,$$

and consider the functions

$$\begin{aligned}
&m_1(i) = \mathbf{E}_i^y x(i), \quad m_1^i(j) = \mathbf{E}_j^y x^i(j), \quad m_2(i) = \mathbf{E}_i^y \eta,\\
&\delta_1(j) = x(j) - m_1(j), \quad \delta_1^i(j) = x^i(j) - m_1^i(j), \quad \delta_2(j) = \eta - m_2(j),\\
&f_{1l}(i,j) = \mathbf{E}_j^y \delta_1^i(j)\delta_l'(j), \quad f_{2k}(j) = \mathbf{E}_j^y \delta_2(j)\delta_k'(j),\\
&K_{lk}(i,j) = \mathbf{E}_i^y \delta_l(i)\delta_k'(j), \quad l,k = 1,2, \quad \gamma(i) = K(i,i),\\
&K(i,j) = \begin{pmatrix} K_{11}(i,j) & K_{12}(i,j)\\ K_{21}(i,j) & K_{22}(i,j)\end{pmatrix}, \quad f(i,j) = \begin{pmatrix} f_{11}(i,j) & f_{12}(i,j)\\ f_{21}(j) & f_{22}(j)\end{pmatrix},\\
&D_1(i,j) = b(i,j)B'(j) + f_{11}(i,j)A'(j), \quad D_2(j) = f_{21}(j)A'(j),\\
&D(i,j) = \begin{pmatrix} D_1(i,j)\\ D_2(j)\end{pmatrix}, \quad m(i) = \begin{pmatrix} m_1(i)\\ m_2(i)\end{pmatrix}.
\end{aligned} \tag{6.7}$$

The solution of the optimal filtering problem (6.4), (6.5) is defined by Theorem 5.2. In compliance with this theorem $m(i)$, $f(i,j)$, $\gamma(i)$ and $K(i,j)$ are defined by the following equations

$$\begin{aligned}
m(i+1) = m(0) &+ \sum_{j=0}^{i} \bar{a}_0(i,j)u(j) + \sum_{j=0}^{i} \bar{a}_1(i,j)m(j)\\
&+ \sum_{j=0}^{i} B(i,j)\Big[y(j+1) - \bar{A}(j)m(j)\Big],
\end{aligned} \tag{6.8}$$

$$\begin{aligned}
f(i,j+1) = \gamma(0) &+ \sum_{l=0}^{j} \Big[A(i,l)f'(j,l) + f(i,l)A'(j,l)\\
&+ A(i,l)\gamma(l)A'(j,l)\Big] + \sum_{l=0}^{j} C(i,l)C'(j,l),
\end{aligned} \tag{6.9}$$

$$\begin{aligned}
\gamma(i+1) = \gamma(0) &+ \sum_{j=0}^{i} \Big[A(i,j)f'(i,j) + f(i,j)A'(i,j)\\
&+ A(i,j)\gamma(j)A'(i,j)\Big] + \sum_{j=0}^{i} C(i,j)C'(i,j),
\end{aligned} \tag{6.10}$$

$$K(i+1,j) = f(i,j+1) + \sum_{l=j+1}^{i} A(i,l)K(l,j), \tag{6.11}$$

where

$$B(i,j) = (D(i,j) + \bar{a}_1(i,j)\gamma(j)\bar{A}'(j))\left[\bar{A}(j)\gamma(j)\bar{A}'(j) + B(j)B'(j)\right]^{\oplus}, \tag{6.12}$$

$$A(i,j) = \bar{a}_1(i,j) - B(i,j)\bar{A}(j), \quad C(i,j) = \bar{b}(i,j) - B(i,j)B(j). \tag{6.13}$$

6.1.2 Solution of the Auxiliary Optimal Control Problem

Transform the first summand in the functional (6.3) by the following way

$$\begin{aligned}
\mathbf{E}x'(N)Fx(N) &= \mathbf{E}\big[\big(x(N) - m_1(N)\big)'Fx(N) + m_1(N)Fx(N)\big) \\
&= \mathbf{E}\big[\big(x(N) - m_1(N)\big)'F\big(x(N) - m_1(N)\big) \\
&\quad + \big(x(N) - m_1(N)\big)'Fm_1(N) + m_1'(N)F\mathbf{E}_N^y x(N)\big] \\
&= \mathbf{E}m_1'(N)Fm_1(N) + Tr\left(F^{1/2}\gamma_{11}(N)F^{1/2}\right).
\end{aligned}$$

Then the functional (6.3) takes the form

$$J(u) = J_1(u) + Tr\left(F^{1/2}\gamma_{11}(N)F^{1/2}\right),$$

where

$$J_1(u) = \mathbf{E}\left[m_1'(N)Fm_1(N) + \sum_{j=0}^{N-1} u'(j)G(j)u(j)\right]. \tag{6.14}$$

Note that in the system of the equations (6.8)–(6.11) the Eq. (6.8) only depends on the control $u(i)$. So, for the solution of the initial optimal control problem (6.1)–(6.3) it is enough to solve the auxiliary optimal control problem (6.8), (6.14). Note also that

$$\begin{aligned}
\bar{a}_1(i,j)\gamma(j)\bar{A}'(j) &= \begin{pmatrix} a_1(i,j) & \tilde{a}_2(j) \\ 0 & 0 \end{pmatrix} \begin{pmatrix} \gamma_{11}(i) & \gamma_{12}(i) \\ \gamma_{21}(i) & \gamma_{22}(i) \end{pmatrix} \begin{pmatrix} A'(i) \\ 0 \end{pmatrix} \\
&= \begin{pmatrix} a_1(i,j) & \tilde{a}_2(j) \\ 0 & 0 \end{pmatrix} \begin{pmatrix} \gamma_{11}(i)A'(i) \\ \gamma_{21}(i)A'(i) \end{pmatrix} \\
&= \begin{pmatrix} (a_1(i,j)\gamma_{11}(i) + \tilde{a}_2(j)\gamma_{21}(i))A'(i) \\ 0 \end{pmatrix}
\end{aligned}$$

From this and (6.6), (6.7), (6.12) it follows that the Eq. (6.8) can be represented in the form

$$m_1(i+1) = m_1(0) + \sum_{j=0}^{i} a_0(i,j)u(j) + \sum_{j=0}^{i} a_1(i,j)m_1(j)$$
$$+ \sum_{j=0}^{i} \tilde{a}_2(j)m_2(j) + \sum_{j=0}^{i} F_1(i,j)[y(j+1) - A(j)m_1(j)], \quad (6.15)$$

$$m_2(i+1) = m_2(0) + \sum_{j=0}^{i} F_2(j)[y(j+1) - A(j)m_1(j)], \quad (6.16)$$

where

$$F_1(i,j) = \left[D_1(i,j) + (a_1(i,j)\gamma_{11}(j) + \tilde{a}_2(j)\gamma_{21}(j))\, A'(j)\right]$$
$$\times [A(j)\gamma_{11}(j)A'(j) + B(j)B'(j)]^{\oplus},$$
$$F_2(j) = D_2(j)[A(j)\gamma_{11}(j)A'(j) + B(j)B'(j)]^{\oplus}. \quad (6.17)$$

Let $u_0(j)$ be the optimal control of the problem (6.8), (6.14), and $v(j)$ be an arbitrary $\mathfrak{F}_j^y$-measurable stochastic process such that $\|v\|^4 < \infty$ in compliance with (2.37). Put

$$u_\varepsilon(i) = u_0(i) + \varepsilon v(i), \quad \varepsilon \geq 0, \quad (6.18)$$

$$J_0(u_0) = \lim_{\varepsilon \to 0} \frac{1}{\varepsilon}\Big[J_1(u_\varepsilon) - J_1(u_0)\Big]. \quad (6.19)$$

Let us calculate the limit (6.19), (6.18) for the optimal control problem (6.15), (6.16), (6.14). Then, solving the equation $J_0(u_0) = 0$, similarly to Theorem 2.2 construct synthesis of the optimal control u_0.

Lemma 6.1 *The limit* (6.19), (6.18) *for the control problem* (6.15), (6.16), (6.14) *there exists and equals*

$$J_0(u_0) = 2\mathbf{E}\left[q'(N)Fm_0^1(N) + \sum_{j=0}^{N-1} v'(j)G(j)u_0(j)\right]. \quad (6.20)$$

Here $q(i)$ is a solution of the equation

$$q(i+1) = \sum_{j=0}^{i} a_0(i,j)v(j) + \sum_{j=0}^{i} a_1(i,j)q(j), \quad q(0) = 0, \quad (6.21)$$

$m_0(i) = \big(m_0^1(i), m_0^2(i)\big)$ is a solution of the system of the equations (6.15), (6.16) *by the control $u = u_0$.*

Proof Put

$$p_\varepsilon^l(i) = \frac{1}{\varepsilon}\Big[m_\varepsilon^l(i) - m_0^l(i)\Big], \quad l = 1, 2, \tag{6.22}$$

where $m_\varepsilon(i) = \big(m_\varepsilon^1(i), m_\varepsilon^2(i)\big)$ is a solution of the system of the equations (6.15), (6.16) by the control (6.18). Using (6.14), (6.18), (6.22), we obtain

$$\begin{aligned}
\frac{1}{\varepsilon}\Big[J_1(u_\varepsilon) - J_1(u_0)\Big] &= \frac{1}{\varepsilon}\mathbf{E}\Big[\big(m_\varepsilon^1(N)\big)' F m_\varepsilon^1(N) - \big(m_0^1(N)\big)' F m_0^1(N) \\
&\quad + \sum_{j=0}^{N-1}\big[u_\varepsilon'(j)G(j)u_\varepsilon(j) - u_0'(j)G(j)u_0(j)\big]\Big] \\
&= \frac{1}{\varepsilon}\mathbf{E}\Big[\big(m_0^1(N) + \varepsilon p_\varepsilon^1(N)\big)' F \big(m_0^1(N) + \varepsilon p_\varepsilon^1(N)\big) \\
&\quad - \big(m_0^1(N)\big)' F m_0^1(N) \\
&\quad + \sum_{j=0}^{N-1}\Big[\big(u_0(j) + \varepsilon v(j)\big)' G(j)\big(u_0(j) + \varepsilon v(j)\big) \\
&\quad - u_0'(j)G(j)u_0(j)\Big]\Big] \\
&= 2\mathbf{E}\Big[\big(p_\varepsilon^1(N)\big)' F m_0^1(N) + \sum_{j=0}^{N-1} v'(j)G(j)u_0(j)\Big] \\
&\quad + \varepsilon\mathbf{E}\Big[\big(p_\varepsilon^1(N)\big)' F p_\varepsilon^1(N) + \sum_{j=0}^{N-1} v'(j)G(j)v(j)\Big].
\end{aligned} \tag{6.23}$$

Let us show that the process $p_\varepsilon^1(i)$ is a solution of the Eq. (6.21), i.e., $p_\varepsilon^1(i) = q(i)$. Let

$$q_\varepsilon(i) = \frac{1}{\varepsilon}\big[x_\varepsilon(i) - x_0(i)\big], \qquad \zeta_\varepsilon(i) = \frac{1}{\varepsilon}\big[y_\varepsilon(i) - y_0(i)\big], \tag{6.24}$$

where $\big(x_\varepsilon(i), y_\varepsilon(i)\big)$ is the solution of the system of the equations (6.15), (6.16) by the control $u_\varepsilon(i)$ (6.18). From (6.1), (6.2), (6.15), and (6.16) it follows that the processes (6.22) and (6.24) are defined by the system of the equations

$$q_\varepsilon(i+1) = \sum_{j=0}^{i} a_0(i,j)v(j) + \sum_{j=0}^{i} a_1(i,j)q_\varepsilon(j), \tag{6.25}$$

$$p_\varepsilon^1(i+1) = \sum_{j=0}^{i} a_0(i,j)v(j) + \sum_{j=0}^{i} a_1(i,j)p_\varepsilon^1(j) + \sum_{j=0}^{i} \tilde{a}_2(j)p_\varepsilon^2(j)$$
$$+ \sum_{j=0}^{i} F_1(i,j)\big[\zeta_\varepsilon(j+1) - A(j)p_\varepsilon^1(j)\big], \tag{6.26}$$

$$p_\varepsilon^2(i+1) = \sum_{j=0}^{i} F_2(j)\big[\zeta_\varepsilon(j+1) - A(j)p_\varepsilon^1(j)\big], \tag{6.27}$$

$$\zeta_\varepsilon(i+1) = A(i)q_\varepsilon(i). \tag{6.28}$$

Substituting (6.27) and (6.28) into (6.26), we have

$$\begin{aligned} p_\varepsilon^1(i+1) &= \sum_{j=0}^{i} a_0(i,j)v(j) + \sum_{j=0}^{i} a_1(i,j)p_\varepsilon^1(j) \\ &\quad + \sum_{j=0}^{i} F_1(i,j)A(j)(q_\varepsilon(j) - p_\varepsilon^1(j)) \\ &\quad + \sum_{j=0}^{i} \tilde{a}_2(j) \sum_{l=0}^{j-1} F_2(l)A(l)\big(q_\varepsilon(l) - p_\varepsilon^1(l)\big) \\ &= \sum_{j=0}^{i} a_0(i,j)v(j) + \sum_{j=0}^{i} a_1(i,j)p_\varepsilon^1(j) + \sum_{j=0}^{i} Z(i,j)\big(q_\varepsilon(j) - p_\varepsilon^1(j)\big), \end{aligned} \tag{6.29}$$

where

$$Z(i,j) = [F_1(i,j) + (a_2(i) - a_2(j+1))F_2(j)]A(j).$$

Subtracting (6.25) from (6.29), we obtain

$$p_\varepsilon^1(i+1) - q_\varepsilon(i+1) = \sum_{j=0}^{i} (a_1(i,j) - Z(i,j))(p_\varepsilon^1(j) - q_\varepsilon(j)).$$

From this and $p_\varepsilon^1(0) = q_\varepsilon(0) = 0$ we obtain $p_\varepsilon^1(i) - q_\varepsilon(i) = 0$, i.e., $p_\varepsilon^1(i) = q_\varepsilon(i)$ for $i \geq 0$. From (6.25), (6.21) it follows also that $q_\varepsilon(i) = q(i)$. So, $p_\varepsilon^1(i) = q(i)$ and via (6.20) one can rewrite (6.23) in the form

$$\frac{1}{\varepsilon}\Big[J_1(u_\varepsilon) - J_1(u_0)\Big] = J_0(u_0) + \varepsilon \mathbf{E}\Big[q'(N)Fq(N) + \sum_{j=0}^{N-1} v'(j)G(j)v(j)\Big].$$

Calculating the limit by $\varepsilon \to 0$, from this we obtain (6.20). The proof is completed.

Theorem 6.1 *Optimal control of the problem* (6.1)–(6.3) *is defined by the expressions*

$$u_0(0) = p(0)\left[\tilde{\psi}_1(N-1,0,I)m_0^1(0) + \psi_1(N-1,0,a_2(\cdot))m_0^2(0)\right], \tag{6.30}$$

$$\begin{aligned} u_0(j+1) &= \alpha_1(j+1)m_0^1(j+1) + \sum_{k=1}^{j} \alpha_2(j+1,k)m_0^1(k) \\ &\quad + \beta_1(j+1)m_0^1(0) + \beta_2(j+1)m_0^2(0) \\ &\quad + \sum_{k=0}^{j} \delta(j+1,k)\left(y_0(k+1) - A(k)m_0^1(k)\right), \quad 0 \le j \le N-1, \end{aligned} \tag{6.31}$$

where

$$\begin{aligned} \alpha_1(j+1) &= p(j+1)\psi_1(N-1,j,I), \\ \alpha_2(j+1,k) &= p(j+1)\psi_1(N-1,j,a_1^j(\cdot,k)) \\ &\quad + Q(j,k)p(k)\psi_1(N-1,k-1,I) \\ &\quad + \sum_{l=k+1}^{j} Q(j,l)p(l)\psi_1(N-1,l-1,a_0^{l-1}(\cdot,k)), \quad 1 \le k \le j, \end{aligned} \tag{6.32}$$

$$\begin{aligned} \beta_1(j+1) &= p(j+1)\psi_1(N-1,j,a_1^j(\cdot,0)) + Q(j,0)p(0)\tilde{\psi}_1(N-1,0,I) \\ &\quad + \sum_{l=1}^{j} Q(j,l)p(l)\psi_1(N-1,l-1,a_0^{l-1}(\cdot,0)), \\ \beta_2(j+1) &= p(j+1)\alpha(N-1,j) + Q(j,0)p(0)\psi_1(N-1,0,a_2(\cdot)) \\ &\quad + \sum_{l=1}^{j} Q(j,l)p(l)\alpha(N-1,l-1), \end{aligned} \tag{6.33}$$

$$\alpha(i,j) = \psi_1(i,j,a_2(\cdot)) - \psi_1(i,j,I)a_2(j), \quad 0 \le j \le i, \tag{6.34}$$

$$\begin{aligned} \delta(j+1,k) &= p(j+1)\left[\psi_1(N-1,j,F_1^j(\cdot,k)) + \alpha(N-1,j)F_2(k)\right] \\ &\quad + \sum_{l=k+1}^{j} Q(j,l)p(l)\left[\psi_1(N-1,l-1,F_1^{l-1}(\cdot,k))\right. \\ &\quad \left. + \alpha(N-1,l-1)F_2(k)\right], \end{aligned} \tag{6.35}$$

$$p(j) = -N_1^{-1}(j)\psi_1'(N-1, j, a_0(\cdot, j))N_0$$
$$\times \left[I + \sum_{k=j}^{N-1} \psi_1(N-1, k, a_0(\cdot, k))N_1^{-1}(k)\psi_1'(N-1, k, a_0(\cdot, k))N_0 \right]^{-1}, \tag{6.36}$$

$$F_1^j(i,k) = F_1(i,k) - F_1(j,k), \quad a_l^j(i,k) = a_l(i,k) - a_l(j,k),$$
$$0 \le k \le j \le i \le N-1, \quad l = 0, 1. \tag{6.37}$$

Here $\psi_1(i, j, f(\cdot))$ for arbitrary function $f(j)$ is defined as

$$\psi_1(i, j, f(\cdot)) = \begin{cases} f(i) + \sum\limits_{k=j+1}^{i} R_1(i,k) f(k-1), & 0 \le j \le i, \\ 0, & j > i, \end{cases} \tag{6.38}$$

and

$$\tilde{\psi}_1(i, 0, I) = I + \sum_{k=0}^{i} R_1(i,k), \tag{6.39}$$

where $R_1(i, j)$ is the resolvent of the kernel $a_1(i, j)$, $Q(j, k)$ is the resolvent of the kernel

$$p(j+1)\psi_1(N-1, j, a_0^j(\cdot, k)), \qquad k \le j \le N-1.$$

Proof Using the resolvent $R_1(i, j)$ of the kernel $a_1(i, j)$, from (6.21) similarly to Lemma 1.3 we obtain

$$\begin{aligned} q(i+1) &= \sum_{j=0}^{i} a_0(i,j)v(j) + \sum_{j=0}^{i} R_1(i,j) \sum_{k=0}^{j-1} a_0(j-1,k)v(k) \\ &= \sum_{j=0}^{i} a_0(i,k)v(k) + \sum_{k=0}^{i} \sum_{j=k+1}^{i} R_1(i,j)a_0(j-1,k)v(k) \\ &= \sum_{j=0}^{i} \psi_1(i, j, a_0(\cdot, j))v(j). \end{aligned}$$

Substituting $q(N)$ into (6.20), we have

$$J_0(u_0) = 2\mathbf{E} \left[\sum_{j=0}^{N-1} v'(j) \left(\psi_1'(N-1, j, a_0(\cdot, j)) F \mathbf{E}_j^{y_0} m_0^1(N) + G(j)u_0(j) \right) \right].$$

From this similarly to Lemma 2.4 and (2.53) it follows that

$$u_0(j) = -G^{-1}(j)\psi_1'(N-1, j, a_0(\cdot, j))F\mathbf{E}_j^{y_0} m_0^1(N), \quad j = 0, 1, \ldots, N-1. \tag{6.40}$$

Let us calculate $\mathbf{E}_j^{y_0} m_0^1(N)$. Substituting (6.16) into (6.15), by virtue of (6.17) and $u = u_0$ we obtain

$$\begin{aligned} m_0^1(i+1) = m_0^1(0) &+ \sum_{j=0}^{i} a_0(i,j)u_0(j) + \sum_{j=0}^{i} a_1(i,j)m_0^1(j) \\ &+ \sum_{j=0}^{i} F_1(i,j)(y_0(j+1) - A(j)m_0^1(j)) \\ &+ \sum_{j=0}^{i} \tilde{a}_2(j)\left[m_0^2(0) + \sum_{l=0}^{j-1} F_2(l)\left(y_0(l+1) - A(l)m_0^1(l)\right)\right]. \end{aligned} \tag{6.41}$$

Note that

$$\begin{aligned} &\sum_{j=0}^{i} \tilde{a}_2(j)\left[m_0^2(0) + \sum_{l=0}^{j-1} F_2(l)\left(y_0(l+1) - A(l)m_0^1(l)\right)\right] \\ &\quad = a_2(i)m_0^2(0) + \sum_{j=0}^{i}\sum_{l=j+1}^{i} \tilde{a}_2(l)F_2(j)\left(y_0(j+1) - A(j)m_0^1(j)\right) \\ &\quad = a_2(i)m_0^2(0) + \sum_{j=0}^{i} (a_2(i) - a_2(j+1))\,F_2(j)\left(y_0(j+1) - A(j)m_0^1(j)\right). \end{aligned}$$

So, (6.41) takes the form

$$\begin{aligned} m_0^1(i+1) = m_0^1(0) + a_2(i)m_0^2(0) &+ \sum_{j=0}^{i} a_0(i,j)u_0(j) + \sum_{j=0}^{i} a_1(i,j)m_0^1(j) \\ &+ \sum_{j=0}^{i} F_1(i,j)\left(y_0(j+1) - A(j)m_0^1(j)\right) \\ &+ \sum_{j=0}^{i} (a_2(i) - a_2(j+1))\,F_2(j)\left(y_0(j+1) - A(j)m_0^1(j)\right). \end{aligned} \tag{6.42}$$

From this it follows that

$$\mathbf{E}_0^{y_0} m_0^1(i+1) = \eta(i+1) + \sum_{j=0}^{i} a_1(i,j)\mathbf{E}_0^{y_0} m_0^1(j), \tag{6.43}$$

where

$$\eta(i) = \begin{cases} m_0^1(0) + a_2(i-1)m_0^2(0) + \sum_{j=0}^{i-1} a_0(i-1,j)\mathbf{E}_0^{y_0} u_0(j), & i \geq 1, \\ m_0^1(0), & i = 0. \end{cases} \quad (6.44)$$

Using the resolvent $R_1(i,l)$ of the kernel $a_1(i,l)$, from (6.43) and (6.44) we obtain

$$\begin{aligned}
\mathbf{E}_0^{y_0} m_0^1(i+1) &= \eta(i+1) + \sum_{j=0}^{i} R_1(i,j)\eta(j) \\
&= \left[I + \sum_{j=0}^{i} R_1(i,j) \right] m_0^1(0) + a_2(i)m_0^2(0) + \sum_{j=0}^{i} a_0(i,j)\mathbf{E}_0^{y_0} u_0(j) \\
&\quad + \sum_{j=1}^{i} R_1(i,j) \left[a_2(j-1)m_0^2(0) + \sum_{l=0}^{j-1} a_0(j-1,l)\mathbf{E}_0^{y_0} u_0(l) \right] \\
&= \tilde{\psi}_1(i,0,I)m_0^1(0) + \psi_1(i,0,a_2(\cdot))m_0^2(0) \\
&\quad + \sum_{l=0}^{i} \psi_1(i,l,a_0(\cdot,l))\mathbf{E}_0^{y_0} u_0(l).
\end{aligned}$$

In particular, for $i = N-1$ we have

$$\mathbf{E}_0^{y_0} m_0^1(N) = \eta_0 + \sum_{l=0}^{N-1} \psi_1(N-1,l,a_0(\cdot,l))\mathbf{E}_0^{y_0} u_0(l), \quad (6.45)$$

where

$$\eta_0 = \tilde{\psi}_1(N-1,0,I)m_0^1(0) + \psi_1(N-1,0,a_2(\cdot))m_0^2(0). \quad (6.46)$$

Substituting (6.40) into (6.45), we obtain

$$\mathbf{E}_0^{y_0} m_0^1(N) = \eta_0 - \sum_{l=0}^{N-1} \psi_1(N-1,l,a_0(\cdot,l))G^{-1}(l)\psi_1'(N-1,l,a_0(\cdot,l))F\mathbf{E}_0^{y_0} m_0^1(N)$$

or

$$\mathbf{E}_0^{y_0} m_0^1(N) = \left[I + \sum_{l=0}^{N-1} \psi_1(N-1,l,a_0(\cdot,l))G^{-1}(l)\psi_1'(N-1,l,a_0(\cdot,l))F \right]^{-1} \eta_0. \quad (6.47)$$

Substituting (6.47) into (6.40) for $j = 0$, we obtain $u_0(0) = p(0)\eta_0$ that coincides with (6.30).

To get (6.31) note that from (6.42) for $i \geq j \geq 0$ we have

$$\begin{aligned}
m_0^1(i+1) - m_0^1(j+1) &= (a_2(i) - a_2(j))m_0^2(0) \\
&\quad + \sum_{l=0}^{i} a_0(i,l)u_0(l) - \sum_{l=0}^{j} a_0(j,l)u_0(l) \\
&\quad + \sum_{l=0}^{i} a_1(i,l)m_0^1(l) - \sum_{l=0}^{j} a_1(j,l)m_0^1(l) \\
&\quad + \sum_{l=0}^{i} F_1(i,l)(y_0(l+1) - A(l)m_0^1(l)) \\
&\quad - \sum_{l=0}^{j} F_1(j,l)(y_0(l+1) - A(l)m_0^1(l)) \\
&\quad + \sum_{l=0}^{i} (a_2(i) - a_2(l+1))F_2(l)(y_0(l+1) - A(l)m_0^1(l)) \\
&\quad - \sum_{l=0}^{j} (a_2(j) - a_2(l+1))F_2(l)(y_0(l+1) - A(l)m_0^1(l)).
\end{aligned}$$

Via (6.37) rewrite the obtained equality in the form

$$\begin{aligned}
m_0^1(i+1) &= m_0^1(j+1) + (a_2(i) - a_2(j))m_0^2(0) \\
&\quad + \sum_{l=0}^{j} a_0^j(i,l)u_0(l) + \sum_{l=j+1}^{i} a_0(i,l)u_0(l) \\
&\quad + \sum_{l=0}^{j} a_1^j(i,l)m_0^1(l) + \sum_{l=j+1}^{i} a_1(i,l)m_0^1(l) \\
&\quad + \sum_{l=0}^{j} \left[F_1^j(i,l) + (a_2(i) - a_2(j))F_2(l)\right]\left(y_0(l+1) - A(l)m_0^1(l)\right) \\
&\quad + \sum_{l=j+1}^{i} \left[F_1(i,l) + (a_2(i) - a_2(l+1))F_2(l)\right]\left(y_0(l+1) - A(l)m_0^1(l)\right).
\end{aligned}$$

Therefore

$$\mathbf{E}_{j+1}^{y_0} m_0^1(i+1) = \zeta(i, j+1) + \sum_{l=j+1}^{i} a_1(i,l)\mathbf{E}_{j+1}^{y_0} m_0^1(l), \tag{6.48}$$

where

$$\begin{aligned}\zeta(i, j+1) &= m_0^1(j+1) + \sum_{l=0}^{j} a_0^j(i,l) u_0(l) + \sum_{l=j+1}^{i} a_0(i,l) \mathbf{E}_{j+1}^{y_0} u_0(l) \\ &+ \sum_{l=0}^{j} a_1^j(i,l) m_0^1(l) + (a_2(i) - a_2(j)) m_0^2(0) \\ &+ \sum_{l=0}^{j} \left[F_1^j(i,l) + (a_2(i) - a_2(j)) F_2(l) \right] \left(y_0(l+1) - A(l) m_0^1(l) \right).\end{aligned}$$

Using the resolvent $R_1(i,l)$ of the kernel $a_1(i,l)$, from (6.48) we have

$$\mathbf{E}_{j+1}^{y_0} m_0^1(i+1) = \zeta(i, j+1) + \sum_{k=j+1}^{i} R_1(i,k) \zeta(k-1, j+1).$$

Substituting $\zeta(i, j+1)$, we obtain

$$\begin{aligned}\mathbf{E}_{j+1}^{y_0} m_0^1(i+1) &= \left[I + \sum_{k=j+1}^{i} R_1(i,k) \right] m_0^1(j+1) \\ &+ \sum_{l=0}^{j} \left[a_0^j(i,l) + \sum_{k=j+1}^{i} R_1(i,k) a_0^j(k-1,l) \right] u_0(l) \\ &+ \sum_{l=j+1}^{i} \left[a_0(i,l) + \sum_{k=j+1}^{i} R_1(i,k) a_0(k-1,l) \right] \mathbf{E}_{j+1}^{y_0} u_0(l) \\ &+ \sum_{l=0}^{j} \left[a_1^j(i,l) + \sum_{k=j+1}^{l} R_1(i,k) a_1^j(k-1,l) \right] m_0^1(l) \\ &+ \left[a_2(i) - a_2(j) + \sum_{k=j+1}^{i} R_1(i,k)(a_2(k-1) - a_2(j)) \right] m_0^2(0) \\ &+ \sum_{l=0}^{j} \Bigg[F_1^j(i,l) + \sum_{k=j+1}^{i} R_1(i,k) F_1^j(k-1,l) \\ &+ \left(a_2(i) - a_2(j) + \sum_{k=j+1}^{i} R_1(i,k)(a_2(k-1) - a_2(j)) \right) F_2(l) \Bigg] \\ &\times \left[y_0(l+1) - A(l) m_0^1(l) \right].\end{aligned}$$

Using the definition (6.38) of the functional $\psi_1(i, j, f(\cdot))$ and (6.34), we have

$$\mathbf{E}^{y_0}_{j+1} m^1_0(i+1) = \psi_1(i, j, I) m^1_0(j+1) + \sum_{l=0}^{j} \psi_1(i, j, a^j_1(\cdot, l)) m^1_0(l) + \alpha(i, j) m^2_0(0)$$
$$+ \sum_{l=0}^{j} \psi_1(i, j, a^j_0(\cdot, l)) u_0(l) + \sum_{l=j+1}^{i} \psi_1(i, l, a_0(\cdot, l)) \mathbf{E}^{y_0}_{j+1} u_0(l)$$
$$+ \sum_{l=0}^{j} (\psi_1(i, j, F^j_1(\cdot, l)) + \alpha(i, j) F_2(l)) \Big[y_0(l+1) - A(l) m^1_0(l) \Big]. \tag{6.49}$$

Putting in (6.49) $i = N - 1$ and

$$\zeta_0(j+1) = \psi_1(N-1, j, I) m^1_0(j+1) + \alpha(N-1, j) m^2_0(0)$$
$$+ \sum_{l=0}^{j} \psi_1(N-1, j, a^j_1(\cdot, l)) m^1_0(l)$$
$$+ \sum_{l=0}^{j} \Big(\psi_1(N-1, j, F^j_1(\cdot, l)) + \alpha(N-1, j) F_2(l) \Big)$$
$$\times [y_0(l+1) - A(l) m^1_0(l)], \quad j \geq 0,$$
$$\zeta_0(0) = \tilde{\psi}_1(N-1, 0, I) m^1_0(0) + \psi_1(N-1, 0, a_2(\cdot)) m^2_0(0), \tag{6.50}$$

rewrite (6.49) in the form

$$\mathbf{E}^{y_0}_{j+1} m^1_0(N) = \zeta_0(j+1) + \sum_{l=0}^{j} \psi_1(N-1, j, a^j_0(\cdot, l)) u_0(l)$$
$$+ \sum_{l=j+1}^{N-1} \psi_1(N-1, l, a_0(\cdot, l)) \mathbf{E}^{y_0}_{j+1} u_0(l). \tag{6.51}$$

From (6.40) for $l > j$ it follows that

$$\mathbf{E}^{y_0}_{j+1} u_0(l) = -G^{-1}(l) \psi'_1(N-1, l, a_0(\cdot, l)) F \mathbf{E}^{y_0}_{j+1} m^1_0(N). \tag{6.52}$$

Substituting (6.52) into (6.51), we obtain

$$\mathbf{E}^{y_0}_{j+1} m^1_0(N) = \left[I + \sum_{l=j+1}^{N-1} \psi_1(N-1, l, a_0(\cdot, l)) G^{-1}(l) \psi'_1(N-1, l, a_0(\cdot, l)) F \right]^{-1}$$
$$\times \left[\zeta_0(j+1) + \sum_{l=0}^{j} \psi_1(N-1, j, a^j_0(\cdot, l)) u_0(l) \right]. \tag{6.53}$$

Substituting (6.53) into (6.40) and using (6.36), we have

$$u_0(j+1) = p(j+1)\zeta_0(j+1) + \sum_{l=0}^{j} p(j+1)\psi_1(N-1, j, a_0^j(\cdot, l))u_0(l).$$

Using the resolvent $Q(j,l)$ of the kernel $p(j+1)\psi_1(N-1, j, a_0^j(\cdot, l))$, represent $u_0(j+1)$ in the form

$$u_0(j+1) = p(j+1)\zeta_0(j+1) + \sum_{k=0}^{j} Q(j,k)p(k)\zeta_0(k). \tag{6.54}$$

Substituting (6.50) into (6.54), we obtain

$$\begin{aligned}
u_0(j+1) &= p(j+1)\zeta_0(j+1) + Q(j,0)p(0)\zeta_0(0) + \sum_{k=1}^{j} Q(j,k)p(k)\zeta_0(k) \\
&= p(j+1)\Bigg[\psi_1(N-1, j, I)m_0^1(j+1) + \alpha(N-1, j)m_0^2(0) \\
&\quad + \sum_{k=0}^{j} \psi_1(N-1, j, a_1^j(\cdot, k))m_0^1(k) \\
&\quad + \sum_{k=0}^{j} \Big(\psi_1(N-1, j, F_1^j(\cdot, k)) + \alpha(N-1, j)F_2(k)\Big) \\
&\quad \times [y_0(k+1) - A(k)m_0^1(k)]\Bigg] \\
&\quad + Q(j,0)p(0)\Big[\tilde{\psi}_1(N-1, 0, I)m_0^1(0) + \psi_1(N-1, 0, a_2(\cdot))m_0^2(0)\Big] \\
&\quad + \sum_{k=1}^{j} Q(j,k)p(k)\Bigg[\psi_1(N-1, k-1, I)m_0^1(k) + \alpha(N-1, k-1)m_0^2(0) \\
&\quad + \sum_{l=0}^{k-1} \psi_1(N-1, k-1, a_0^{k-1}(\cdot, l))m_0^1(l) \\
&\quad + \sum_{l=0}^{k-1} \Big(\psi_1(N-1, k-1, F_1^{k-1}(\cdot, l)) + \alpha(N-1, k-1)F_2(l)\Big) \\
&\quad \times [y_0(l+1) - A(l)m_0^1(l)]\Bigg].
\end{aligned}$$

Note that

$$\sum_{k=1}^{j} Q(j,k)p(k)\Big[\sum_{l=0}^{k-1}\psi_1(N-1,k-1,a_0^{k-1}(\cdot,l))m_0^1(l)$$

$$+\sum_{l=0}^{k-1}\Big(\psi_1(N-1,k-1,F_1^{k-1}(\cdot,l))$$

$$+\alpha(N-1,k-1)F_2(l)\Big)[y_0(l+1)-A(l)m_0^1(l)]\Big]$$

$$=\sum_{l=0}^{j}\sum_{k=l+1}^{j} Q(j,k)p(k)\psi_1(N-1,k-1,a_0^{k-1}(\cdot,l))m_0^1(l)$$

$$+\sum_{l=0}^{j}\sum_{k=l+1}^{j} Q(j,k)p(k)\Big(\psi_1(N-1,k-1,F_1^{k-1}(\cdot,l))$$

$$+\alpha(N-1,k-1)F_2(l)\Big)[y_0(l+1)-A(l)m_0^1(l)].$$

So, $u_0(j+1)$ takes the form

$$u_0(j+1)=p(j+1)\psi_1(N-1,j,I)m_0^1(j+1)$$

$$+\sum_{l=1}^{j}\Big[p(j+1)\psi_1(N-1,j,a_1^j(\cdot,l))+Q(j,l)p(l)\psi_1(N-1,l-1,I)$$

$$+\sum_{k=l+1}^{j} Q(j,k)p(k)\psi_1(N-1,k-1,a_0^{k-1}(\cdot,l))\Big]m_0^1(l)$$

$$+\Big[p(j+1)\psi_1(N-1,j,a_1^j(\cdot,0))+Q(j,0)p(0)\tilde{\psi}_1(N-1,0,I)$$

$$+\sum_{k=1}^{j} Q(j,k)p(k)\psi_1(N-1,k-1,a_0^{k-1}(\cdot,0))\Big]m_0^1(0)$$

$$+\Big[p(j+1)\alpha(N-1,j)+Q(j,0)p(0)\psi_1(N-1,0,a_2(\cdot))$$

$$+\sum_{l=0}^{j} Q(j,l)p(l)\alpha(N-1,l-1)\Big]m_0^2(0)$$

$$+\sum_{l=0}^{j}\Big[p(j+1)\Big(\psi_1(N-1,j,F_1^j(\cdot,l))+\alpha(N-1,j)F_2(l)\Big)$$

$$+\sum_{k=l+1}^{j} Q(j,k)p(k)\Big(\psi_1(N-1,k-1,F_1^{k-1}(\cdot,l))$$
$$+\alpha(N-1,k-1)F_2(l)\Big)\Big]$$
$$\times [y_0(l+1)-A(l)m_0^1(l)],$$

that via (6.32)–(6.35) coincides with (6.31). Theorem is proved.

Theorem 6.2 *Let $m_0^1(0) = m_0^2(0) = 0$. Then the optimal trajectory $m_0(i) = (m_0^1(i), m_0^2(i))$ and the optimal control $u_0(i)$ of the problem* (6.15), (6.16), *and* (6.14) *can be represented in the form*

$$m_0^l(i+1) = \sum_{j=0}^{i} G_0^l(i,j)y_0(j+1), \qquad l=1,2, \tag{6.55}$$

$$u_0(i+1) = \sum_{j=0}^{i} Q_0(i,j)y_0(j+1), \tag{6.56}$$

where

$$G_0^1(i,j) = \psi_0(i,j,L(\cdot,j)),$$
$$L(i,j) = F_1(i,j) + \sum_{k=j+1}^{i} a_0(i,k)\delta(k,j) + (a_2(i)-a_2(j+1))F_2(j),$$
$$G_0^2(i,j) = F_2(j) - \sum_{k=j+1}^{i} F_2(k)A(k)G_0^1(k-1,j), \tag{6.57}$$

for arbitrary function $f(\cdot)$ the functional $\psi_0(i,j,f(\cdot))$ is defined by the formula

$$\psi_0(i,j,f(\cdot)) = f(i) + \sum_{l=j+1}^{i} R_0(i,l)f(l-1), \tag{6.58}$$

where $R_0(i,j)$ is the resolvent of the kernel

$$S_0(i,j) = a_0(i,j)\alpha_1(j) + a_1(i,j) - F_1(i,j)A(j)$$
$$+\sum_{k=j+1}^{i} a_0(i,k)[\alpha_2(k,j) - \delta(k,j)A(j)]$$
$$-(a_2(i)-a_2(j+1))F_2(j)A(j), \tag{6.59}$$

and

$$Q_0(i,j) = \delta(i+1,j) + \alpha_1(i+1)G_0^1(i,j) + \sum_{k=j+1}^{i} [\alpha_2(i+1,k) - \delta(i+1,k)A(k)]G_0^1(k-1,j), \qquad (6.60)$$

$\alpha(j)$, $\alpha_2(i,j)$, $\delta(i,j)$ *and* $F_1(i,j)$, $F_2(j)$ *are defined by* (6.34), (6.32), (6.35), *and* (6.17), *respectively.*

Proof Note that via $m_0^1(0) = m_0^2(0) = 0$ from (6.30) we have $u_0(0) = 0$. Substituting (6.16) and (6.31) into (6.15) for $m^l(i) = m_0^l(i)$, $l = 1, 2$, $u(i) = u_0(i)$, $y(i) = y_0(i)$ and using (6.57), (6.59), we obtain

$$\begin{aligned}
m_1(i+1) &= \sum_{j=1}^{i} a_0(i,j)\Big[\alpha_1(j)m_0^1(j) + \sum_{k=1}^{j-1}\alpha_2(j,k)m_0^1(k) \\
&\quad + \sum_{k=0}^{j-1}\delta(j,k)\left(y_0(k+1) - A(k)m_0^1(k)\right)\Big] \\
&\quad + \sum_{k=1}^{i} a_1(i,k)m_1(k) + \sum_{j=0}^{i}\tilde{a}_2(j)\sum_{l=0}^{j-1}F_2(l)[y_0(l+1) - A(l)m_0^1(l)] \\
&\quad + \sum_{j=0}^{i}F_1(i,j)[y_0(j+1) - A(j)m_0^1(j)] \\
&= \sum_{k=1}^{i}\Big[a_0(i,k)\alpha_1(k) + a_1(i,k) - F_1(i,k)A(k) \\
&\quad + \sum_{j=k+1}^{i} a_0(i,j)[\alpha_2(j,k) - \delta(j,k)A(k)] \\
&\quad - (a_2(i) - a_2(k+1))F_2(k)A(k)\Big]m_0^1(k) \\
&\quad + \sum_{k=0}^{i}\Big[F_1(i,k) + \sum_{j=k+1}^{i} a_0(i,j)\delta(j,k) \\
&\quad + (a_2(i) - a_2(k+1))F_2(k)\Big]y_0(k+1) \\
&= \sum_{k=0}^{i} L(i,k)y_0(k+1) + \sum_{k=1}^{i} S_0(i,k)m_0^1(k).
\end{aligned}$$

Using the resolvent $R_0(i,j)$ of the kernel $S_0(i,j)$, via (6.58), (6.57) we get

$$
\begin{aligned}
m_0^1(i+1) &= \sum_{k=0}^{i} L(i,k)y_0(k+1) + \sum_{j=1}^{i} R_0(i,j) \sum_{k=0}^{j-1} L(j-1,k)y_0(k+1) \\
&= \sum_{k=0}^{i} L(i,k)y_0(k+1) + \sum_{k=0}^{i} \sum_{j=k+1}^{i} R_0(i,j)L(j-1,k)y_0(k+1) \\
&= \sum_{k=0}^{i} \psi_0(i,k,L(\cdot,k))y_0(k+1) \\
&= \sum_{k=0}^{i} G_0^1(i,k)y_0(k+1). \qquad (6.61)
\end{aligned}
$$

Substituting (6.61) into (6.16) and using $y = y_0$, $m_2(0) = 0$ and (6.57), we have

$$
\begin{aligned}
m_0^2(i+1) &= \sum_{j=0}^{i} F_2(j)y_0(j+1) - \sum_{j=0}^{i} F_2(j)A(j) \sum_{k=0}^{j-1} G_0^1(j-1,k)y_0(k+1) \\
&= \sum_{k=0}^{i} \left[F_2(k) - \sum_{j=k+1}^{i} F_2(j)A(j)G_0^1(j-1,k) \right] y_0(k+1) \\
&= \sum_{k=0}^{i} G_0^2(i,k)y_0(k+1).
\end{aligned}
$$

Substituting (6.61) into (6.31), we obtain

$$
\begin{aligned}
u_0(j+1) &= \alpha_1(j+1) \sum_{l=0}^{j} G_0^1(j,l)y_0(l+1) + \sum_{k=1}^{j} \alpha_2(j+1,k) \\
&\quad \times \sum_{l=0}^{k-1} G_0^1(k-1,l)y_0(l+1) \\
&\quad + \sum_{k=0}^{j} \delta(j+1,k) \left[y_0(k+1) - A(k) \sum_{l=0}^{k-1} G_0^1(k-1,l)y_0(l+1) \right] \\
&= \sum_{l=0}^{j} \Bigg[\alpha_1(j+1)G_0^1(j,l) + \delta(j+1,l) \\
&\quad + \sum_{k=l+1}^{j} [\alpha_2(j+1,k) - \delta(j+1,k)A(k)]G_0^1(k-1,l) \Bigg] y_0(l+1) \\
&= \sum_{k=0}^{j} Q_0(j,k)y_0(k+1).
\end{aligned}
$$

The proof is completed.

Corollary 6.1 *Let the conditions of Theorem* 6.1 *hold and* $a_2(i) = 0$. *Then* $m_0^2(i) = 0$.

Proof Via (6.55), (6.57), (6.17) it is enough to show that $D_2(i) = 0$. Since $a_2(i) = 0$ then $x(i)$ and η are mutually independent. So, via (6.7) we have

$$\begin{aligned} D_2(j) &= f_{21}(j)A'(j) \\ &= \mathbf{E}_j^y \delta_2(j)\delta_1'(j)A'(j) \\ &= \mathbf{E}_j^y \delta_2(j)\mathbf{E}_j^y \delta_1'(j)A'(j) \\ &= 0. \end{aligned}$$

The proof is completed.

Corollary 6.2 *Let the conditions of Theorem* 6.1 *hold and*

$$a_0(i, j) = 0, \qquad a_2(i) = 0. \tag{6.62}$$

Therefore, the system (6.1) *is uncontrolled and the problem of optimal filtering only is considered. By that the representation* (6.61) *has the form*

$$m_0^1(i + 1) = \sum_{j=0}^{i} G_0^1(i, j)y(j + 1),$$

where

$$\begin{aligned} G_0^1(i, j) &= \psi_0(i, j, F_1(\cdot, j)), \\ F_1(i, j) &= \left[D_1(i, j) + a_1(i, j)\gamma_{11}(j)A'(j)\right] \\ &\quad \times \left[A(j)\gamma_{11}(j)A'(j) + B(j)B'(j)\right]^{\oplus}. \end{aligned} \tag{6.63}$$

Proof By the conditions (6.62) the Eq. (6.1) takes the form

$$x(i + 1) = x_0 + \sum_{j=0}^{i} a_1(i, j)x(j) + \sum_{j=0}^{i} b(i, j)\xi(j + 1)$$

and from (6.57), (6.17) it follows that $L(i, j) = F_1(i, j)$ and (6.63). The proof is completed.

6.2 The Method of Integral Representations

In the previous section it was shown that the solution of the optimal control problem is reduced (in each moment of time) to the solution of the system of four stochastic difference equations, i.e., to another hard enough problem.

In this section, another way to the solution of the considered optimal control problem is proposed: the system of deterministic equations is obtained that define the matrices $G_0^l(i,j), l=1,2$, and $Q_0(i,j)$ for the representations (6.55) and (6.56). This method allows to construct the optimal control immediately from observations, provided that the matrices $G_0^l(i,j), l=1,2$, and $Q_0(i,j)$ are defined previously.

Consider once again the optimal control problem for the process (6.1) with the performance functional (6.3) and observable process

$$y(i+1)=A(i)x(i-h)+B(i)\xi(i+1),\quad y(0)=0,\quad x(j)=0,\quad j<0. \tag{6.64}$$

Lemma 6.2 *By the assumptions of Theorem* 6.2, *the optimal control of the problem* (6.1), (6.64), (6.3) *is represented in the form*

$$u_0(i+1)=\sum_{j=0}^{i}Q_0(i,j)y_0(j+1),\quad i=0,\dots,N-1, \tag{6.65}$$

where $Q_0(i,j)$ is a deterministic matrix, $y_0(j)$ is a solution of the system (6.1), (6.64).

Proof Similarly to Chap. 2, it can be established that the limit (6.19), (6.18) for the control problem (6.1), (6.64), (6.3) there exists and equals

$$J_0(u_0)=2\mathbf{E}\left[q'(N)Fx_0(N)+\sum_{j=0}^{N-1}v'(j)G(j)u_0(j)\right],$$

where

$$q(i+1)=\sum_{j=0}^{i}a_0(i,j)v(j)+\sum_{j=0}^{i}a_1(i,j)q(j),\quad q(0)=0.$$

Using the resolvent $R_1(i,j)$ of the kernel $a_1(i,j)$ and (6.38), similarly to (2.53), (6.40), we obtain

$$u_0(i)=-G^{-1}(i)\psi_1'(N-1,i,a_0(\cdot,i))F\mathbf{E}_i^{y_0}x_0(N). \tag{6.66}$$

Note that via (6.7) $\mathbf{E}_i^{y_0}x_0(N)=\mathbf{E}_i^{y_0}\mathbf{E}_N^{y_0}x_0(N)=\mathbf{E}_i^{y_0}m_0(N)$. So, (6.66) coincides with (6.40) and via (6.55) the mean square estimates $m_0^l(i)$ are representable in the form

$$m_0^l(i+1)=\sum_{j=0}^{i}G_0^l(i,j)y_0(j+1),\quad l=1,2,\quad i=0,\ldots,N-1. \tag{6.67}$$

The relations (6.66) and (6.67) imply (6.65) for

$$Q_0(i,j)=-G^{-1}(i+1)\psi_1'(N-1,i+1,a_0(\cdot,i+1))FG_0^1(i,j),$$
$$0\le j\le i\le N-1. \tag{6.68}$$

The lemma is proved.

From the proof of Lemma 6.2 it follows that the matrices $Q_0(i,j)$ and $G_0^1(i,j)$ that define the optimal control (6.65) and the optimal estimate (6.67), are connected by (6.68). Let us get another equation that connects these matrices.

Put

$$q_0(i,j)=\begin{cases}\sum\limits_{k=j+1}^{i}a_0(i,k)Q_0(k-1,j), & i>j,\\ 0, & i\le j,\end{cases} \tag{6.69}$$

and

$$P_0(i,j)=\begin{cases}a_1(i,j)+q_0(i,j+h)A(j+h), & 0\le j<i-h,\\ a_1(i,j), & i-h\le j\le i,\end{cases} \tag{6.70}$$

$$P_1(i,j)=b(i,j)+q_0(i,j)B(j),\quad j\le i. \tag{6.71}$$

Let $R_0(i,j)$ be the resolvent of the kernel $P_0(i,j)$,

$$S_1(i,j)=R(i+1,j-h)A'(j)+\psi_0(i,j,P_1(\cdot,j))B'(j), \tag{6.72}$$

$$S_2(j)=F(j-h-1)A'(j),\quad F(i)=\gamma_1\psi_0'(i,0,a_2(\cdot)), \tag{6.73}$$

$$\begin{aligned}K(l,j)&=A(l)R(l-h,j-h)A'(j)\\&\quad+A(l)\psi_0(l-h-1,j,P_1(\cdot,j))B'(j)\\&\quad+B(l)\psi_0'(j-h-1,l,P_1(\cdot,l))A'(j),\quad l,j\ge 0,\end{aligned} \tag{6.74}$$

$$\begin{aligned}R(i+1,j+1)&=\tilde{\psi}_0(i,0,I)\gamma_0\tilde{\psi}_0'(j,0,I)\\&\quad+\psi_0(i,0,a_2(\cdot))\gamma_1\psi_0'(j,0,a_2(\cdot))\\&\quad+\sum_{l=0}^{i\wedge j}\psi_0(i,l,P_1(\cdot,l))\psi_0'(j,l,P_1(\cdot,l)),\quad i,j\ge 0,\end{aligned} \tag{6.75}$$

$$R(i,j)=\begin{cases}\tilde{\psi}_0(i-1,0,I)\gamma_0, & i\geq 1,\ j=0,\\ \gamma_0, & i=j=0,\\ 0, & i\wedge j<0,\end{cases}$$

$\psi_0(i,j,f(\cdot))$ for arbitrary function $f(j)$ and $\tilde{\psi}_0(i,0,I)$ are defined as

$$\psi_0(i,j,f(\cdot))=\begin{cases}f(i)+\sum\limits_{l=j+1}^{i}R_0(i,l)f(l-1), & i\geq j,\\ 0, & i<j,\end{cases}$$

$$\tilde{\psi}_0(i,0,I)=I+\sum_{l=0}^{i}R_0(i,l),\quad i\geq 0. \tag{6.76}$$

Theorem 6.3 *The matrices $Q_0(i,j)$ and $G_0^l(i,j)$, $l=1,2$, in the representations* (6.65), (6.67) *are connected by the equations*

$$\begin{aligned}G_0^1(i,j)B(j)B'(j)&=S_1(i,j)-\sum_{l=0}^{i}G_0^1(i,l)K(l,j),\\ G_0^2(i,j)B(j)B'(j)&=S_2(j)-\sum_{l=0}^{i}G_0^2(i,l)K(l,j),\\ 0\leq j\leq i,&\quad i=0,\ldots,N-1,\end{aligned} \tag{6.77}$$

where $S_1(i,j)$, $S_2(j)$, $K(l,j)$ are defined by the conditions (6.69)–(6.76).

Proof Let $m_l(i)$, $l=1,2$ be $\mathfrak{F}_i^{y_0}$-measurable stochastic processes of the form (6.67) with arbitrary kernels $G_l(i,j)$. It is easy to see that

$$\mathbf{E}\begin{pmatrix}x_0(i)-m_0^1(i)\\ \eta-m_0^2(i)\end{pmatrix}\begin{pmatrix}m_1(i)\\ m_2(i)\end{pmatrix}'=0,$$

or

$$\begin{aligned}\mathbf{E}x_0(i)m_l'(i)&=\mathbf{E}m_0^1(i)m_l'(i),\\ \mathbf{E}\eta m_l'(i)&=\mathbf{E}m_0^2(i)m_l'(i).\end{aligned} \tag{6.78}$$

Calculate the left-hand side of the first equality (6.78)

$$\begin{aligned}\mathbf{E}x_0(i+1)m_l'(i+1)&=\mathbf{E}x_0(i+1)\sum_{j=0}^{i}[A(j)x_0(j-h)+B(j)\xi(j+1)]'G_l'(i,j)\\ &=\sum_{j=0}^{i}\Big[\mathbf{E}\big(x_0(i+1)x_0'(j-h)\big)A'(j)\\ &\quad+\mathbf{E}\big(x_0(i+1)\xi'(j+1)\big)B'(j)\Big]G_l'(i,j)\end{aligned} \tag{6.79}$$

and the right-hand side of this first equality

$$\begin{aligned}\mathbf{E}m_0^1(i+1)m_l'(i+1) &= \mathbf{E}\sum_{l=0}^{i} G_0^1(i,l)[A(l)x_0(l-h)+B(l)\xi(l+1)]\\ &\quad\times\sum_{j=0}^{i}[A(j)x_0(j-h)+B(j)\xi(j+1)]'G_l'(i,j)\\ &= \sum_{j=0}^{i}\sum_{l=0}^{i} G_0^1(i,l)\Big[A(l)\mathbf{E}\big(x_0(l-h)x_0'(j-h)\big)A'(j)\\ &\quad + A(l)\mathbf{E}\big(x_0(l-h)\xi'(j+1)\big)B'(j)\\ &\quad + B(l)\mathbf{E}\big(\xi(l+1)x_0'(j-h)\big)A'(j)\Big]G_l'(i,j)\\ &\quad + \sum_{j=0}^{i} G_0^1(i,j)B(j)B'(j)G_l'(i,j).\end{aligned} \tag{6.80}$$

Equating (6.79) and (6.80) and using that $G_l(i,j)$ is an arbitrary matrix, from (6.78)–(6.80) we have

$$\begin{aligned}&\mathbf{E}\big(x_0(i+1)x_0'(j-h)\big)A'(j)+\mathbf{E}\big(x_0(i+1)\xi'(j+1)\big)B'(j)\\ &\quad= \sum_{l=0}^{i} G_0^1(i,l)\Big[A(l)\mathbf{E}\big(x_0(l-h)x_0'(j-h)\big)A'(j)\\ &\quad + A(l)\mathbf{E}\big(x_0(l-h)\xi'(j+1)\big)\times B'(j)\\ &\quad + B(l)\mathbf{E}\big(\xi(l+1)x_0'(j-h)\big)A'(j)\Big] + G_0^1(i,j)B(j)B'(j).\end{aligned} \tag{6.81}$$

To calculate expectations in (6.81) substitute (6.56) into (6.1). Then

$$\begin{aligned}x_0(i+1) &= x_0 + \sum_{j=0}^{i} a_0(i,j)\sum_{k=0}^{j-1} Q_0(j-1,k)y_0(k+1)\\ &\quad + \sum_{j=0}^{i} a_1(i,j)x_0(j) + a_2(i)\eta + \sum_{j=0}^{i} b(i,j)\xi(j+1)\\ &= x_0 + a_2(i)\eta + \sum_{j=0}^{i} a_1(i,j)x_0(j) + \sum_{j=0}^{i} b(i,j)\xi(j+1)\\ &\quad + \sum_{k=0}^{i}\sum_{j=k+1}^{i} a_0(i,j)Q_0(j-1,k)y_0(k+1).\end{aligned} \tag{6.82}$$

Substitute (6.64), (6.69) in the last summand

$$\sum_{k=0}^{i}\sum_{j=k+1}^{i} a_0(i,j)Q_0(j-1,k)y_0(k+1) = \sum_{k=0}^{i} q_0(i,k)(A(k)x_0(k-h) + B(k)\xi(k+1))$$

and note that via (6.64)

$$\sum_{k=0}^{i} q_0(i,k)A(k)x_0(k-h) = \sum_{k=h}^{i} q_0(i,k)A(k)x_0(k-h) = \sum_{j=0}^{i-h} q_0(i,j+h)A(j+h)x_0(j).$$

As a result using (6.70), (6.71), rewrite (6.82) in the form

$$x_0(i+1) = \zeta(i+1) + \sum_{j=0}^{i} P_0(i,j)x_0(j),$$
$$\zeta(i+1) = x_0 + a_2(i)\eta + \sum_{j=0}^{i} P_1(i,j)\xi(j+1), \quad i \geq 0, \qquad \zeta(0) = x_0.$$

Using the resolvent $R_0(i,j)$ of the kernel $P_0(i,j)$ and (6.76), from this we obtain

$$\begin{aligned} x_0(i+1) &= \zeta(i+1) + \sum_{j=0}^{i} R_0(i,j)\zeta(j) \\ &= \zeta(i+1) + R_0(i,j)x_0 \\ &\quad + \sum_{j=1}^{i} R_0(i,j)\left(x_0 + a_2(j-1)\eta + \sum_{k=0}^{j-1} P_1(j-1,k)\xi(k+1)\right) \\ &= \tilde{\psi}_0(i,0,I)x_0 + \psi_0(i,0,a_2(\cdot))\eta + \sum_{j=0}^{i} \psi_0(i,j,P_1(\cdot,j))\xi(j+1), \quad i \geq 0. \end{aligned} \tag{6.83}$$

From (6.73), (6.75), (6.83) it follows that

$$\begin{aligned} \mathbf{E}\big(x_0(i)x_0'(j)\big) &= R(i,j), \quad i,j \geq 1, \\ \mathbf{E}\big(x_0(i+1)\xi'(j+1)\big) &= \psi_0(i,j,P_1(\cdot,j)), \quad i \geq j \geq 0, \\ \mathbf{E}\big(\eta x_0'(i+1)\big) &= \gamma_1\psi_0'(i,0,a_2(\cdot)) = F(i), \quad i \geq 0. \end{aligned} \tag{6.84}$$

Substituting (6.84) into (6.81), we obtain the first equality (6.77).

Transforming similarly the second equality (6.78) to the form of (6.81), we obtain

$$\begin{aligned}
\mathbf{E}(\eta x_0'(j-h))A'(j) = \sum_{l=0}^{i} G_0^2(i,l)\Big[& A(l)\mathbf{E}\big(x_0(l-h)x_0'(j-h)\big)A'(j) \\
& + A(l)\mathbf{E}(x_0(l-h)\xi'(j+1))B'(j) \\
& + B(l)\mathbf{E}(\xi(l+1)x_0'(j-h))A'(j)\Big] \\
& + G_0^2(i,j)B(j)B'(j).
\end{aligned}$$

From this via (6.73)–(6.75), (6.83) the second equality (6.77) follows. The theorem is proved.

The system of the deterministic equations (6.68), (6.77) defined the matrices $Q_0(i,j)$ and $G_0^l(i,j), l=1,2$, can be solved previously. It allows us to construct the optimal control u_0 and the optimal estimates m_0^1, m_0^2 immediately from observations. For simplicity let us consider some scalar examples.

Example 6.1 Consider the problem (6.1), (6.2) and (6.3) with $N=1$. From (6.68)–(6.77) for $h=0$ it follows that $Q_0(0,0)=0$, and therefore $u_0(1)=0$. Besides of that

$$m_0^1(1) = G_0^1(0,0)y_0(1), \quad m_0^2(1) = G_0^2(0,0)y_0(1), \tag{6.85}$$

where

$$\begin{aligned}
G_0^1(0,0) &= S_1(0,0)[K(0,0)+B(0)B'(0)]^{\oplus}, \\
G_0^2(0,0) &= S_2(0)[K(0,0)+B(0)B'(0)]^{\oplus}.
\end{aligned} \tag{6.86}$$

To define $G_0^1(0,0)$, $G_0^2(0,0)$ in the terms of the initial parameters note that

$$\begin{gathered}
q_0(0,0)=0, \quad P_0(0,0)=a_1(0,0), \quad P_1(0,0)=b(0,0), \\
S_1(0,0)=R(1,0)A'(0)+b(0,0)B'(0), \\
R(1,0)=(I+a_1(0,0))\gamma_0, \quad K(0,0)=A(0)\gamma_0 A'(0).
\end{gathered} \tag{6.87}$$

So,

$$S_1(0,0)=(I+a_1(0,0))\gamma_0 A'(0)+b(0,0)B'(0), \quad S_2(0)=0,$$

and via (6.85)–(6.87)

$$\begin{aligned}
m_0^1(1) &= [(I+a_1(0,0))\gamma_0 A'(0)+b(0,0)B'(0)] \\
&\quad \times [A(0)\gamma_0 A'(0)+B(0)B'(0)]^{\oplus} y_0(1), \\
m_0^2(1) &= 0.
\end{aligned}$$

Example 6.2 Consider the scalar problem (6.1), (6.2), (6.3) with $N=2$. By assumption that $m_0^1(0)=m_0^2(0)=0$ we have $u_0(0)=0$ and

$$\begin{aligned} u_0(1) &= Q_0(0,0)y_0(1), \\ m_0^1(1) = G_0^1(0,0)y_0(1), \quad & m_0^1(2) = G_0^1(1,0)y_0(1) + G_0^1(1,1)y_0(2), \\ m_0^2(1) = G_0^2(0,0)y_0(1), \quad & m_0^2(2) = G_0^2(1,0)y_0(1) + G_0^2(1,1)y_0(2). \end{aligned} \tag{6.88}$$

To define $Q_0(0,0)$ note that via (6.68), (6.38)

$$\begin{aligned} Q_0(0,0) &= -\frac{F\psi_1(1,1,a_0(\cdot,1))G_0^1(0,0)}{G(1)} \\ &= -\frac{Fa_0(1,1)G_0^1(0,0)}{G(1)}. \end{aligned} \tag{6.89}$$

From (6.77) it follows that

$$\begin{aligned} G_0^1(0,0)(B^2(0)+K(0,0)) &= S_1(0,0), \\ G_0^1(0,0)(B^2(0)+K(0,0)) &= S_2(0), \end{aligned} \tag{6.90}$$

where via (6.72), (6.74), (6.75)

$$\begin{aligned} S_1(0,0) = R(1,0)A(0) + b(0,0)B(0), \quad S_2(0) = 0, \\ R(1,0) = (1+a_1(0,0))\gamma_0, \quad K(0,0) = A^2(0)\gamma_0. \end{aligned} \tag{6.91}$$

As a result we obtain $G_0^1(0,0)$, $G_0^2(0,0)$ and $Q_0(0,0)$ in the terms of the initial parameters of the considered problem

$$G_0^1(0,0) = \frac{(1+a_1(0,0))A(0)\gamma_0 + b(0,0)B(0)}{A^2(0)\gamma_0 + B^2(0)}, \quad G_0^2(0,0) = 0, \tag{6.92}$$

and

$$Q_0(0,0) = -\frac{Fa_0(1,1)[(1+a_1(0,0))A(0)\gamma_0 + b(0,0)B(0)]}{G(1)[A^2(0)\gamma_0 + B^2(0)]}. \tag{6.93}$$

From (6.77) it follows that the equations for $G_0^1(1,0)$, $G_0^1(1,1)$, and $G_0^2(1,0)$, $G_0^2(1,1)$ are, respectively,

$$\begin{aligned} G_0^1(1,0)(B^2(0)+K(0,0)) + G_0^1(1,1)K(1,0) &= S_1(1,0), \\ G_0^1(1,0)K(0,1) + G_0^1(1,1)(B^2(1)+K(1,1)) &= S_1(1,1), \end{aligned}$$

and

$$\begin{aligned} G_0^2(1,0)(B^2(0)+K(0,0)) + G_0^2(1,1)K(1,0) &= S_2(0), \\ G_0^2(1,0)K(0,1) + G_0^2(1,1)(B^2(1)+K(1,1)) &= S_2(1), \end{aligned}$$

with the solutions

$$
\begin{aligned}
G_0^1(1,0) &= \frac{S_1(1,0)(B^2(1)+K(1,1)) - S_1(1,1)K(1,0)}{(B^2(0)+K(0,0))(B^2(1)+K(1,1)) - K^2(1,0)},\\
G_0^1(1,1) &= \frac{S_1(1,1)(B^2(0)+K(0,0)) - S_1(1,0)K(1,0)}{(B^2(0)+K(0,0))(B^2(1)+K(1,1)) - K^2(1,0)},
\end{aligned} \tag{6.94}
$$

and

$$
\begin{aligned}
G_0^2(1,0) &= \frac{S_2(0)(B^2(1)+K(1,1)) - S_2(1)K(1,0)}{(B^2(0)+K(0,0))(B^2(1)+K(1,1)) - K^2(1,0)},\\
G_0^2(1,1) &= \frac{S_2(1)(B^2(0)+K(0,0)) - S_2(0)K(1,0)}{(B^2(0)+K(0,0))(B^2(1)+K(1,1)) - K^2(1,0)},
\end{aligned} \tag{6.95}
$$

where via (6.72)–(6.74)

$$
\begin{aligned}
S_1(1,0) &= R(2,0)A(0) + \psi_0(1,0,P_1(\cdot,0))B(0),\\
S_1(1,1) &= R(2,1)A(1) + \psi_0(1,1,P_1(\cdot,1))B(0),\\
S_2(0) &= 0, \quad S_2(1) = \gamma_1\psi_0(0,0,a_2(\cdot))A(1),\\
K(0,0) &= R(0,0)A^2(0), \qquad K(1,1) = R(1,1)A^2(1),\\
K(1,0) &= [A(0)R(1,0) + B(0)\psi_0(0,0,P_1(\cdot,0))]A(1).
\end{aligned} \tag{6.96}
$$

To calculate $S(i,j)$, $K(i,j)$ note that via (6.75)

$$
\begin{aligned}
R(0,0) &= \gamma_0,\\
R(1,0) &= \bar{\psi}_0(0,0,1)\gamma_0,\\
R(1,1) &= \bar{\psi}_0^2(0,0,1)\gamma_0 + \psi_0^2(0,0,a_2(\cdot))\gamma_1 + \psi_0^2(0,0,P_1(\cdot,0)),\\
R(2,0) &= \bar{\psi}_0(1,0,1)\gamma_0,\\
R(2,1) &= \bar{\psi}_0(1,0,1)\bar{\psi}_0'(0,0,1)\gamma_0 + \psi_0(1,0,a_2(\cdot))\psi_0'(0,0,a_2(\cdot))\gamma_1\\
&\quad + \psi_0(1,0,P_1(\cdot,0))\psi_0'(0,0,P_1(\cdot,0)).
\end{aligned} \tag{6.97}
$$

So, using (6.76), we define

$$
\begin{aligned}
\bar{\psi}_0(0,0,1) &= 1 + R_0(0,0),\\
\bar{\psi}_0(1,0,1) &= 1 + R_0(1,0) + R_0(1,1),\\
\psi_0(0,0,a_2(\cdot)) &= a_2(0),\\
\psi_0(1,0,a_2(\cdot)) &= a_2(1) + R_0(1,1)a_2(0),\\
\psi_0(0,0,P_1(\cdot,0)) &= P_1(0,0),\\
\psi_0(1,0,P_1(\cdot,0)) &= P_1(1,0) + R_0(1,1)P_1(0,0),\\
\psi_0(1,1,P_1(\cdot,0)) &= P_1(1,0).
\end{aligned} \tag{6.98}
$$

Now we have to define $P_1(i, j)$ and the resolvent $R_0(i, j)$ of the kernel $P_0(i, j)$. Note that via (6.69), we have

$$
\begin{aligned}
q_0(0,0) &= 0,\\
q_0(1,0) &= a_0(1,1)Q_0(0,0),\\
q_0(1,1) &= 0.
\end{aligned}
\tag{6.99}
$$

So, from (6.70), (6.71), and (6.99) we obtain

$$
\begin{aligned}
P_0(0,0) &= a_1(0,0),\\
P_0(1,0) &= a_1(1,0) + q_0(1,0)A(0)\\
&= a_1(1,0) + a_0(1,1)Q_0(0,0)A(0),\\
P_0(1,1) &= a_1(1,1),\\
P_1(0,0) &= b(0,0),\\
P_1(1,0) &= b(1,0) + q_0(1,0)B(0)\\
&= b(1,0) + a_0(1,1)Q_0(0,0)B(0),\\
P_1(1,1) &= b(1,1).
\end{aligned}
\tag{6.100}
$$

Using the definition of the resolvent (1.18) and (6.100), we have

$$
\begin{aligned}
R_0(0,0) &= P_0(0,0) = a_1(0,0),\\
R_0(1,1) &= P_0(1,1) = a_1(1,1),\\
R_0(1,0) &= P_0(1,0) + P_0(1,1)P_0(0,0)\\
&= a_1(1,0) + a_0(1,1)Q_0(0,0)A(0) + a_1(1,1)a_1(0,0),
\end{aligned}
\tag{6.101}
$$

where $Q_0(0, 0)$ is defined in (6.93).

Substituting (6.101), (6.100) into (6.98) and then sequently into (6.97), (6.96), (6.95), (6.94) we obtain $G_0^l(1, 0)$, $G_0^l(1, 1)$ and therefore the estimates (6.85) in the terms of the initial parameters.

6.3 Quasilinear Stochastic Difference Volterra Equation

Consider the optimal control problem for the quasilinear stochastic difference Volterra equation

$$
\begin{aligned}
x(i+1) &= x_0 + \sum_{j=0}^{i} a_0(i,j)u(j) + \sum_{j=0}^{i} a_1(i,j)x(j)\\
&\quad + a_2(i)\eta + \varepsilon\Phi(i+1, x_{i+1}) + \sum_{j=0}^{i} b(i,j)\xi(j+1)
\end{aligned}
\tag{6.102}
$$

under the observations (6.64) with the performance functional (6.3).

Let us suppose that the functional $\Phi(i,\varphi)$ depends on $\varphi(j)$ for $j=0,1,\dots,i$ only and satisfies the conditions: $\Phi(i,\varphi)\in\mathbf{R}^n$, $\Phi(i,0)=0$,

$$\big|\Phi(i,\varphi)\big|\le\sum_{j=0}^{i}\big(1+|\varphi(j)|\big)K_1(j), \tag{6.103}$$

$$\big|\Phi(i,\varphi_1)-\Phi(i,\varphi_2)\big|\le\sum_{j=0}^{i}\big|\varphi_1(j)-\varphi_2(j)\big|K_1(j), \tag{6.104}$$

for arbitrary functions $\varphi,\varphi_1,\varphi_2\in\tilde{H}$.

Let $\big(x_\varepsilon^u,y_\varepsilon^u\big)$ be the solution of the system of the equations (6.102), (6.64) under a control $u\in\mathbf{U}$, $\varepsilon\ge 0$; $J_\varepsilon(u)$ be the functional (6.3) for $x=x_\varepsilon^u$, $V_\varepsilon=\inf\limits_{u\in\mathbf{U}}J_\varepsilon(u)$.

Note that the control problem (6.102), (6.64), (6.3) for $\varepsilon=0$ coincides with the control problem (6.1), (6.64), (6.3) for which the optimal control u_0 is known and is defined by (6.65).

Theorem 6.3 *The control $u_0(j)$ defined by* (6.65) *is the zeroth approximation to the optimal control of the control problem* (6.102), (6.64), (6.3).

Proof Let $\rho_\varepsilon=\sup\limits_{u\in\mathbf{U}}\big|J_\varepsilon(u)-J_0(u)\big|$ defines the closeness of two control problems (6.1), (6.64), (6.3) and (6.102), (6.64), (6.3). Via $u_0\in\mathbf{U}$ and $J_0(u_0)=V_0$ we have (Lemma 1.1)

$$0\le J_\varepsilon(u_0)-V_\varepsilon\le|J_\varepsilon(u_0)-J_0(u_0)|+|V_0-V_\varepsilon|\le 2\rho_\varepsilon. \tag{6.105}$$

So, via Definition 3.1 to prove the theorem it is enough to show that $\rho_\varepsilon\le C\varepsilon$. Using (6.3) and $\|x_\varepsilon^u\|^2<\infty$, we obtain

$$\begin{aligned}|J_\varepsilon(u)-J_0(u)|&=\mathbf{E}\big|(x_\varepsilon^u(N))'Fx_\varepsilon^u(N)-(x_0^u(N))'Fx_0^u(N)\big|\\&=\mathbf{E}|(x_\varepsilon^u(N)-x_0^u(N))'F(x_\varepsilon^u(N)+x_0^u(N))|\\&\le|F|\sqrt{\mathbf{E}|x_\varepsilon^u(N)-x_0^u(N)|^2\mathbf{E}|x_\varepsilon^u(N)+x_0^u(N)|^2}\\&\le C\|x_\varepsilon^u-x_0^u\|_N.\end{aligned} \tag{6.106}$$

Let us prove that $\|x_\varepsilon^u-x_0^u\|_N\le C\varepsilon$. From (6.102), (6.103) it follows that

$$|x_\varepsilon^u(i+1)-x_0^u(i+1)|\le\varepsilon\sum_{j=0}^{i+1}(1+|x_\varepsilon^u(j)|)K_1(j)+\sum_{j=1}^{i}|a_1(i,j)||x_\varepsilon^u(j)-x_0^u(j)|.$$

Squaring the obtained inequality, calculating expectation and putting

$$z(j) = \mathbf{E}\left|x_\varepsilon^u(j) - x_0^u(j)\right|^2,$$

we obtain

$$z(i+1) \le C\left[\varepsilon^2 + \sum_{j=1}^{i} z(j)\right].$$

From this, using Lemma 1.2, we obtain $\|x_\varepsilon^u - x_0^u\|_N \le C\varepsilon$ that via (6.106) implies $\rho_\varepsilon \le C\varepsilon$. The theorem is proved.

Put

$$u_\varepsilon(i+1) = \sum_{j=0}^{i} Q_0(i,j) y_\varepsilon^u(j+1), \tag{6.107}$$

where $Q_0(i,j)$ is defined by the system of the equations (6.69)–(6.76). Note that for $\varepsilon = 0$ (6.106) coincides with (6.65).

Theorem 6.4 *The control* (6.107) *is the zeroth approximation to the optimal control of the control problem* (6.102), (6.64), (6.3) *for* $\varepsilon > 0$.

Proof To prove the inequality $0 \le J_\varepsilon(u_\varepsilon) - V_\varepsilon \le C\varepsilon$ note that

$$0 \le J_\varepsilon(u_\varepsilon) - V_\varepsilon \le |J_\varepsilon(u_\varepsilon) - J_\varepsilon(u_0)| + |J_\varepsilon(u_0) - V_\varepsilon|. \tag{6.108}$$

Via (6.105) and the proof of Theorem 6.3 we have $|J_\varepsilon(u_0) - V_\varepsilon| \le C\varepsilon$. So, from (6.108) and (6.3) it follows that for some $C > 0$

$$0 \le J_\varepsilon(u_\varepsilon) - V_\varepsilon \le C(\varepsilon + \|u_\varepsilon - u_0\|_N + \|x_\varepsilon^{u_\varepsilon} - x_\varepsilon^{u_0}\|_N). \tag{6.109}$$

From (6.107), (6.65), (6.64) we have

$$\begin{aligned}
\mathbf{E}\left|u_\varepsilon(i+1) - u_0(i+1)\right|^2 &= \mathbf{E}\left|\sum_{j=0}^{i} Q_0(i,j)\left[y_\varepsilon^{u_\varepsilon}(j+1) - y_0^{u_0}(j+1)\right]\right|^2 \\
&= \mathbf{E}\left|\sum_{j=h+1}^{i} Q_0(i,j)A(j)\left[x_\varepsilon^{u_\varepsilon}(j-h) - x_0^{u_0}(j-h)\right]\right|^2 \\
&\le C\sum_{j=h+1}^{i} \mathbf{E}|x_\varepsilon^{u_\varepsilon}(j-h) - x_0^{u_0}(j-h)|^2 \\
&\le C\|x_\varepsilon^{u_\varepsilon} - x_0^{u_0}\|_i^2.
\end{aligned} \tag{6.110}$$

Via (6.102)

$$x_\varepsilon^{u_\varepsilon}(i+1) - x_0^{u_0}(i+1) = \sum_{j=0}^{i} a_0(i,j)[u_\varepsilon(j) - u_0(j)] + \sum_{j=1}^{i} a_1(i,j)\Big[x_\varepsilon^{u_\varepsilon}(j) - x_0^{u_0}(j)\Big] + \varepsilon\Phi(i+1, x_{\varepsilon,i+1}^{u_\varepsilon}).$$

Squaring this equality and calculating expectation, via (6.103) we obtain

$$\mathbf{E}|x_\varepsilon^{u_\varepsilon}(i+1) - x_0^{u_0}(i+1)|^2 \le C\left[\varepsilon^2 + \sum_{j=0}^{i} \mathbf{E}|u_\varepsilon(j) - u_0(j)|^2 + \sum_{j=1}^{i} \mathbf{E}|x_\varepsilon^{u_\varepsilon}(j) - x_0^{u_0}(j)|^2\right].$$

From this via (6.110) it follows that the following inequality holds

$$\mathbf{E}|x_\varepsilon^{u_\varepsilon}(i+1) - x_0^{u_0}(i+1)|^2 \le C\left[\varepsilon^2 + \sum_{j=1}^{i} \|x_\varepsilon^{u_\varepsilon} - x_0^{u_0}\|_j^2\right].$$

Via Lemma 1.2 we obtain $\|x_\varepsilon^{u_\varepsilon} - x_0^{u_0}\|_N \le C\varepsilon$ and therefore via (6.110)

$$\|u_\varepsilon - u_0\|_N \le C\varepsilon. \tag{6.111}$$

So, via (6.109) it is enough to prove that $\|x_\varepsilon^{u_\varepsilon} - x_\varepsilon^{u_0}\|_N \le C\varepsilon$. From (6.102) it follows that

$$|x_\varepsilon^{u_\varepsilon}(i+1) - x_\varepsilon^{u_0}(i+1)| \le \sum_{j=0}^{i} |a_0(i,j)||u_\varepsilon(j) - u_0(j)| + \sum_{j=1}^{i} |a_1(i,j)||x_\varepsilon^{u_\varepsilon}(j) - x_\varepsilon^{u_0}(j)| + \varepsilon|\Phi(i+1, x_{\varepsilon,i+1}^{u_\varepsilon}) - \Phi(i+1, x_{\varepsilon,i+1}^{u_0})|.$$

Squaring and calculating the expectation, via (6.104) and (6.111) for $z(i) = \mathbf{E}|x_\varepsilon^{u_\varepsilon}(i) - x_\varepsilon^{u_0}(i)|^2$ we obtain

$$\begin{aligned} z(i+1) &\le C\left[\sum_{j=0}^{i} \mathbf{E}|u_\varepsilon(j) - u_0(j)|^2 + \sum_{j=1}^{i} z(j) + \varepsilon^2 \sum_{j=1}^{i+1} z(j)\right] \\ &\le C\left[\varepsilon^2 + \sum_{j=1}^{i} z(j) + \varepsilon^2 z(i+1)\right]. \end{aligned}$$

For small enough ε ($C\varepsilon^2 < 1$) from this it follows that

$$z(i+1) \le \frac{C}{1 - C\varepsilon^2}\left[\varepsilon^2 + \sum_{j=1}^{i} z(j)\right]$$

and via Lemma 1.2 $\|z\|_N \le C\varepsilon^2$ for some $C > 0$. The proof is completed.

6.4 Linear Stochastic Difference Equation with Unknown Parameter

In this section, another way of the optimal control construction is proposed. The optimal control in final form is obtained for the optimal control problem for stochastic linear difference equation with unknown parameter and quadratic performance functional. Numerical calculations illustrate and continue the theoretical investigations.

6.4.1 Statement of the Problem

Consider the optimal control problem for stochastic linear difference equation

$$x(i+1) = a_0(i)u(i) + \sum_{j=0}^{i} a_1(i-j)x(j) + a_2(i)\eta + b(i)\xi(i+1) \qquad (6.112)$$

and the quadratic performance functional

$$J(u) = \mathbf{E}\left[x'(N)Fx(N) + \sum_{j=0}^{N-1}\left(x'(j)F_1(j)x(j) + u'(j)G(j)u(j)\right)\right]. \qquad (6.113)$$

Here $\xi(i) \in \mathbf{R}^m$, $i = 1, \ldots, N$, are Gaussian mutually independent random variables such that $\mathbf{E}\xi(i) = 0$, $\mathbf{E}\xi(i)\xi'(i) = I$, I is the identity matrix, $\eta \in \mathbf{R}^r$ is unknown Gaussian parameter, $u(i) \in \mathbf{R}^l$ is a control, F, $F_1(j)$, $G(j)$ are positive semidefinite and $a_0(i)$, $a_1(i)$, $a_2(i)$ are arbitrary matrices with corresponding dimensions.

The problem is to find a control $u_0(i)$ for which the performance functional $J(u)$ is minimal, i.e., $J(u_0) = \inf_{u \in U} J(u)$, U is the set of admissible control.

Let $\mathfrak{F}_i$ be σ-algebra induced by the values of $x(j)$, $j = 1, 2, \ldots, i$. It is known [122] that the optimal (in the mean square sense) estimate of the unknown parameter

η is defined by the conditional expectation $m(i) = \mathbf{E}\{\eta/\mathfrak{F}_i\}$. This estimate $m(i)$ is defined by the system of two equations

$$m(i+1) = m(i) + W(i)\left[x(i+1) - a_0(i)u(i) - \sum_{j=0}^{i} a_1(i-j)x(j) - a_2(i)m(i)\right], \tag{6.114}$$

$$\gamma(i+1) = \gamma(i) - W(i)a_2(i)\gamma(i).$$

Here

$$W(i) = \gamma(i)a_2'(i)[b(i)b'(i) + a_2(i)\gamma(i)a_2'(i)]^+, \tag{6.115}$$

A^+ is the pseudoinverse matrix of the matrix A and

$$\gamma(i) = \mathbf{E}\{(\eta - m(i))(\eta - m(i))'/\mathfrak{F}_i\}. \tag{6.116}$$

So, the optimal control problem (6.112), (6.113) reduces to the optimal control problem (6.112)–(6.115).

Theorem 6.5 *Let there exist a nonnegative functional* $V(i, x_i, y(i)) = V(i, x(0), \dots, x(i), y(i))$ *and a control* $u_0(i, x_i, y(i)) = u_0(i, x(0), \dots, x(i), y(i))$ *such that*

$$\begin{aligned} &\inf_{u\in U} \mathbf{E}[\Delta V(i, x_i^u, m^u(i)) + (x^u(i))' F_1(i) x^u(i) + u'(i)G(i)u(i)] \\ &= \mathbf{E}[\Delta V(i, x_i^0, m^0(i)) + (x^0(i))' F_1(i) x^0(i) + u_0'(i)G(i)u_0(i)] \\ &= 0, \end{aligned} \tag{6.117}$$

$$V(N, x_N, y(N)) = x'(N)Fx(N), \tag{6.118}$$

where $x^u(i)$ *and* $m^u(i)$ *is the solution of the system* (6.112), (6.114) *with the control* $u(i)$, $x^0(i)$ *and* $m^0(i)$ *is the solution of the system* (6.112), (6.114) *with the control* $u_0(i)$,

$$\Delta V(i, x_i^u, m^u(i)) = V(i+1, x_{i+1}^u, m^u(i+1)) - V(i, x_i^u, m^u(i)). \tag{6.119}$$

Then $u_0(i)$ *is the optimal control of the control problem* (6.112), (6.113) *and*

$$J(u_0) = \mathbf{E}V(0, x(0), m(0)). \tag{6.120}$$

Proof From (6.117) it follows that for $u \in U$

$$\begin{aligned} &\mathbf{E}[\Delta V(i, x_i^u, m^u(i)) + (x^u(i))' F_1(i) x^u(i) + u'(i)G(i)u(i)] \\ &\geq \mathbf{E}[\Delta V(i, x_i^0, m^0(i)) + (x^0(i))' F_1(i) x^0(i) + u_0'(i)G(i)u_0(i)] \\ &= 0, \end{aligned}$$

Summing the obtained inequality over $i = 0, 1, \dots, N-1$, we obtain

$$\begin{aligned}
&\sum_{i=0}^{N-1} \mathbf{E}[\Delta V(i, x_i^u, m^u(i)) + (x^u(i))' F_1(i) x^u(i) + u'(i) G(i) u(i)] \\
&\quad = \mathbf{E}V(N, x_N^u, m^u(N)) - \mathbf{E}V(0, x(0), m(0)) \\
&\qquad + \sum_{i=0}^{N-1} \mathbf{E}[(x^u(i))' F_1(i) x^u(i) + u'(i) G(i) u(i)] \\
&\quad \geq \sum_{i=0}^{N-1} \mathbf{E}[\Delta V(i, x_i^0, m^0(i)) + (x^0(i))' F_1(i) x^0(i) + u_0'(i) G(i) u_0(i)] \\
&\quad = \mathbf{E}V(N, x_N^0, m^0(N)) - \mathbf{E}V(0, x(0), m(0)) \\
&\qquad + \sum_{i=0}^{N-1} \mathbf{E}[(x^0(i))' F_1(i) x^0(i) + u_0'(i) G(i) u_0(i)] \\
&\quad = 0.
\end{aligned}$$

From this and (6.113), (6.118) it follows that

$$J(u) \geq \mathbf{E}V(0, x(0), m(0)) = J(u_0).$$

The proof is completed.

6.4.2 The Optimal Control Construction

Below we will construct the functional $V(i, x_i, y(i))$ satisfying the conditions of Theorem 6.5 and the optimal control $u_0(i, x_i, y(i))$ in some final form.

We will construct the functional $V(i) = V(i, x_i, y(i))$ satisfying the conditions (6.116) and (6.117) in the form

$$\begin{aligned}
V(i) &= x'(i) P_0(i) x(i) + 2x'(i) P_1(i) y(i) + y'(i) P_2(i) y(i) + P_3(i) \\
&\quad + 2\sum_{j=0}^{i-1} x'(i) Q_0(i, j) x(j) + 2\sum_{j=0}^{i-1} y'(i) Q_1(i, j) x(j) \\
&\quad + \sum_{j=0}^{i-1} \sum_{k=0}^{i-1} x'(j) R(i, j, k) x(k)
\end{aligned} \tag{6.121}$$

Here it is assumed that $P_0(i)$, $P_1(i)$, $P_2(i)$ are matrices of dimensions $n\times n, n\times r, r\times r$, respectively, $P_0'(i) = P_0(i)$, $P_2'(i) = P_2(i)$, $R'(i, j, k) = R(i, k, j)$, $P_3(i) \geq 0$, the matrix

$$\begin{pmatrix} P_0(i) & P_1(i) \\ P_1'(i) & P_2(i) \end{pmatrix}$$

is positive semidefinite. It is assumed also that $\sum_{j=0}^{-1} = 0$ and

$$\begin{gathered} P_0(N) = F, \\ P_1(N) = 0, \quad P_2(N) = 0, \quad P_3(N) = 0, \\ R(N, j, k) = 0, \quad Q_0(N, j) = 0, \quad Q_1(N, j) = 0. \end{gathered} \tag{6.122}$$

To calculate $\mathbf{E}\Delta V(i) = \mathbf{E}\Delta V(i, x_i, m(i))$ for the solution of the equations (6.112), (6.114) note that via (6.117) we have

$$\mathbf{E}\Delta V(i) = \alpha_1(i) + 2\alpha_2(i) + \alpha_3(i) + 2\alpha_4(i) + 2\alpha_5(i) + \alpha_6(i) + \Delta P_3(i), \tag{6.123}$$

where

$$\begin{aligned} \alpha_1(i) &= \mathbf{E}[x'(i+1)P_0(i+1)x(i+1) - x'(i)P_0(i)x(i)], \\ \alpha_2(i) &= \mathbf{E}[x'(i+1)P_1(i+1)m(i+1) - x'(i)P_1(i)m(i)], \\ \alpha_3(i) &= \mathbf{E}[m'(i+1)P_2(i+1)m(i+1) - m'(i)P_2(i)m(i)], \\ \alpha_4(i) &= \mathbf{E}\left[\sum_{j=0}^{i} x'(i+1)Q_0(i+1, j)x(j) - \sum_{j=0}^{i-1} x'(i)Q_0(i, j)x(j)\right], \\ \alpha_5(i) &= \mathbf{E}\left[\sum_{j=0}^{i} m'(i+1)Q_1(i+1, j)x(j) - \sum_{j=0}^{i-1} m'(i)Q_1(i, j)x(j)\right], \\ \alpha_6(i) &= \mathbf{E}\left[\sum_{j=0}^{i}\sum_{k=0}^{i} x'(j)R(i+1, j, k)x(k) - \sum_{j=0}^{i-1}\sum_{k=0}^{i-1} x'(j)R(i, j, k)x(k)\right]. \end{aligned} \tag{6.124}$$

Via (6.124), (6.112) transform $\alpha_1(i)$ by the following way

$$\begin{aligned} \alpha_1(i) = \mathbf{E}\Bigg[\Bigg(& a_0(i)u(i) + \sum_{j=0}^{i} a_1(i-j)x(j) \\ & + a_2(i)(\eta - m(i)) + a_2(i)m(i) + b(i)\xi(i+1)\Bigg)' P_0(i+1) \end{aligned}$$

$$\times \left(a_0(i)u(i) + \sum_{k=0}^{i} a_1(i-k)x(k) + a_2(i)(\eta - m(i)) + a_2(i)m(i) \right.$$
$$\left. \left. + b(i)\xi(i+1) \right) - x'(i)P_0(i)x(i) \right]$$
$$= \mathbf{E}\left[u'(i)a_0'(i)P_0(i+1)a_0(i)u(i) + m'(i)a_2'(i)P_0(i+1)a_2(i)m(i) \right.$$
$$+ \sum_{j=0}^{i}\sum_{k=0}^{i} x'(j)a_1'(i-j)P_0(i+1)a_1(i-k)x(k)$$
$$+ 2\sum_{k=0}^{i} m'(i)a_2'(i)P_0(i+1)a_1(i-k)x(k)$$
$$+ 2\sum_{k=0}^{i} u'(i)a_0'(i)P_0(i+1)a_1(i-k)x(k)$$
$$+ Tr[a_2'(i)P_0(i+1)a_2(i)\gamma(i) + b'(i)P_0(i+1)b(i)]$$
$$\left. + 2u'(i)a_0'P_0(i+1)a_2(i)m(i) - x'(i)P_0(i)x(i) \right]. \tag{6.125}$$

Note also that

$$\sum_{j=0}^{i}\sum_{k=0}^{i} x'(j)a_1'(i-j)P_0(i+1)a_1(i-k)x(k)$$
$$= \left(\sum_{j=0}^{i-1} x'(j)a_1'(i-j) + x'(i)a_1'(0) \right) P_0(i+1) \left(\sum_{k=0}^{i-1} a_1(i-k)x(k) + a_1(0)x(i) \right)$$
$$= \sum_{j=0}^{i-1}\sum_{k=0}^{i-1} x'(j)a_1'(i-j)P_0(i+1)a_1(i-k)x(k)$$
$$+ 2\sum_{j=0}^{i-1} x'(j)a_1'(i-j)P_0(i+1)a_1(0)x(i) + x'(i)a_1'(0)P_0(i+1)a_1(0)x(i) \tag{6.126}$$

and

$$\sum_{k=0}^{i} m'(i)a_2'(i)P_0(i+1)a_1(i-k)x(k)$$
$$= \sum_{k=0}^{i-1} m'(i)a_2'(i)P_0(i+1)a_1(i-k)x(k) + m'(i)a_2'(i)P_0(i+1)a_1(0)x(i). \tag{6.127}$$

Substituting (6.126), (6.127) into (6.125), as a result we obtain

$$\begin{aligned}
\alpha_1(i) = \mathbf{E}\Bigg[& x'(i)\big(a_1'(0)P_0(i+1)a_1(0) - P_0(i)\big)x(i) \\
& + 2x'(i)a_1'(0)P_0(i+1)a_2(i)m(i) + m'(i)a_2'(i)P_0(i+1)a_2(i)m(i) \\
& + Tr[a_2'(i)P_0(i+1)a_2(i)\gamma(i) + b'(i)P_0(i+1)b(i)] \\
& + 2\sum_{j=0}^{i-1} x'(i)a_1'(0)P_0(i+1)a_1(i-j)x(j) \\
& + 2\sum_{k=0}^{i-1} m'(i)a_2'(i)P_0(i+1)a_1(i-k)x(k) \\
& + \sum_{j=0}^{i-1}\sum_{k=0}^{i-1} x'(j)a_1'(i-j)P_0(i+1)a_1(i-k)x(k) \\
& + 2u'(i)a_0'(i)P_0(i+1)\left(a_2(i)m(i) + \sum_{k=0}^{i} a_1(i-k)x(k)\right) \\
& + u'(i)a_0'(i)P_0(i+1)a_0(i)u(i)\Bigg]. \qquad (6.128)
\end{aligned}$$

To calculate $\alpha_2(i)$ note that via (6.112), (6.114), we have

$$m(i+1) = m(i) + W(i)[a_2(i)(\eta - m(i)) + b(i)\xi(i+1)]. \qquad (6.129)$$

Therefore, from (6.124), (6.112), (6.129), we obtain

$$\begin{aligned}
\alpha_2(i) = \mathbf{E}\Bigg[& \Bigg(a_0(i)u(i) + \sum_{j=0}^{i} a_1(i-j)x(j) \\
& + a_2(i)(\eta - m(i)) + a_2(i)m(i) + b(i)\xi(i+1)\Bigg)' P_1(i+1) \\
& \times \big(m(i) + W(i)[a_2(i)(\eta - m(i)) + b(i)\xi(i+1)]\big) - x'(i)P_1(i)m(i)\Bigg] \\
= \mathbf{E}\Bigg[& u'(i)a_0'(i)P_1(i+1)m(i) + \sum_{j=0}^{i} x'(j)a_1'(i-j)P_1(i+1)m(i) \\
& + Tr[a_2'(i)P_1(i+1)W(i)a_2(i)\gamma(i) + b'(i)P_1(i+1)W(i)b(i)] \\
& + m'(i)a_2'(i)P_1(i+1)m(i) - x'(i)P_1(i)m(i)\Bigg].
\end{aligned}$$

Using the equality

$$\sum_{j=0}^{i} x'(j)a_1'(i-j) = x'(i)a_1'(0) + \sum_{j=0}^{i-1} x'(j)a_1'(i-j),$$

represent $\alpha_2(i)$ in the form

$$\begin{aligned}\alpha_2(i) = \mathbf{E}\Big[& x'(i)(a_1'(0)P_1(i+1) - P_1(i))m(i) + m'(i)a_2'(i)P_1(i+1)m(i) \\ & + Tr[a_2'(i)P_1(i+1)W(i)a_2(i)\gamma(i) + b'(i)P_1(i+1)W(i)b(i)] \\ & + \sum_{j=0}^{i-1} x'(j)a_1'(i-j)P_1(i+1)m(i) + u'(i)a_0'(0)P_1(i+1)m(i)\Big].\end{aligned} \tag{6.130}$$

Similarly, via (6.124), (6.129) we have

$$\begin{aligned}\alpha_3(i) &= \mathbf{E}\big[\big(m(i) + W(i)[a_2(i)(\eta - m(i)) + b(i)\xi(i+1)]\big)P_2(i+1) \\ &\quad \times \big(m(i) + W(i)[a_2(i)(\eta - m(i)) + b(i)\xi(i+1)]\big) - m'(i)P_2(i)m(i)\big] \\ &= \mathbf{E}\big[m'(i)\Delta P_2(i)m(i) + Tr[a_2'(i)W'(i)P_2(i+1)W(i)a_2(i)\gamma(i) \\ &\quad + b'(i)W'(i)P_2(i+1)W(i)b(i)]\big].\end{aligned} \tag{6.131}$$

Calculating $\alpha_4(i)$, via (6.124), (6.112) we obtain

$$\begin{aligned}\alpha_4(i) = \mathbf{E}\Bigg[& \sum_{j=0}^{i} \Big(a_0(i)u(i) + \sum_{k=0}^{i} a_1(i-k)x(k) + a_2(i)(\eta - m(i)) + a_2(i)m(i) \\ & + b(i)\xi(i+1)\Big)' Q_0(i+1,j)x(j) - \sum_{j=0}^{i-1} x'(i)Q_0(i,j)x(j)\Bigg] \\ = \mathbf{E}\Bigg[& \sum_{j=0}^{i}\sum_{k=0}^{i} x'(k)a_1'(i-k)Q_0(i+1,j)x(j) \\ & + \sum_{j=0}^{i} u'(i)a_0'(i)Q_0(i+1,j)x(j) + \sum_{j=0}^{i} m'(i)a_2'(i)Q_0(i+1,j)x(j) \\ & - \sum_{j=0}^{i-1} x'(i)Q_0(i,j)x(j)\Bigg].\end{aligned}$$

Similarly to (6.126) we have

$$\begin{aligned}
&\sum_{j=0}^{i}\sum_{k=0}^{i} x'(k)a_1'(i-k)Q_0(i+1,j)x(j) \\
&\quad = \left(\sum_{k=0}^{i-1} x'(k)a_1'(i-k) + x'(i)a_1'(0)\right) \\
&\qquad \times \left(\sum_{j=0}^{i-1} Q_0(i+1,j)x(j) + Q_0(i+1,i)x(i)\right) \\
&\quad = \sum_{j=0}^{i-1}\sum_{k=0}^{i-1} x'(k)a_1'(i-k)Q_0(i+1,j)x(j) + \sum_{j=0}^{i-1} x'(i)a_1'(0)Q_0(i+1,j)x(j) \\
&\qquad + \sum_{k=0}^{i-1} x'(k)a_1'(i-k)Q_0(i+1,i)x(i) + x'(i)a_1'(0)Q_0(i+1,i)x(i).
\end{aligned}$$

So,

$$\begin{aligned}
\alpha_4(i) = \mathbf{E}\Bigg[& x'(i)a_1'(0)Q_0(i+1,i)x(i) + m'(i)a_2'(i)Q_0(i+1,i)x(i) \\
& + \sum_{j=0}^{i-1} x'(i)[a_1'(0)Q_0(i+1,j) + Q_0'(i+1,i)a_1(i-j) - Q_0(i,j)]x(j) \\
& + \sum_{j=0}^{i-1} m'(i)a_2'(i)Q_0(i+1,j)x(j) \\
& + \sum_{j=0}^{i-1}\sum_{k=0}^{i-1} x'(k)a_1'(i-k)Q_0(i+1,j)x(j) \\
& + u'(i)a_0'(i)\sum_{j=0}^{i} Q_0(i+1,j)x(j)\Bigg]
\end{aligned} \tag{6.132}$$

and similarly

$$\begin{aligned}
\alpha_5(i) = \mathbf{E}\Bigg[& \sum_{j=0}^{i}\Big(m(i) + W(i)[a_2(i)(\eta - m(i)) + b(i)\xi(i+1)]\Big)' \\
& \times Q_1(i+1,j)x(j) - \sum_{j=0}^{i-1} m'(i)Q_1(i,j)x(j)\Bigg]
\end{aligned}$$

$$
\begin{aligned}
&= \mathbf{E}\left[\sum_{j=0}^{i} m'(i)Q_1(i+1,j)x(j) - \sum_{j=0}^{i-1} m'(i)Q_1(i,j)x(j)\right] \\
&= \mathbf{E}\left[m'(i)Q_1(i+1,i)x(i) + \sum_{j=0}^{i-1} m'(i)\Delta Q_1(i,j)x(j)\right]. \qquad (6.133)
\end{aligned}
$$

To calculate $\alpha_6(i)$ note that

$$
\begin{aligned}
&\sum_{j=0}^{i}\sum_{k=0}^{i} x'(j)R(i+1,j,k)x(k) \\
&\quad = \sum_{j=0}^{i}\left(\sum_{k=0}^{i-1} x'(j)R(i+1,j,k)x(k) + x'(j)R(i+1,j,i)x(i)\right) \\
&\quad = \sum_{j=0}^{i-1}\left(\sum_{k=0}^{i-1} x'(j)R(i+1,j,k)x(k) + x'(j)R(i+1,j,i)x(i)\right) \\
&\qquad + \sum_{k=0}^{i-1} x'(i)R(i+1,i,k)x(k) + x'(i)R(i+1,i,i)x(i) \\
&\quad = x'(i)R(i+1,i,i)x(i) + 2\sum_{j=0}^{i-1} x'(i)R(i+1,i,j)x(j) \\
&\qquad + \sum_{j=0}^{i-1}\sum_{k=0}^{i-1} x'(j)R(i+1,j,k)x(k).
\end{aligned}
$$

So,

$$
\begin{aligned}
\alpha_6(i) = \mathbf{E}\Bigg[& x'(i)R(i+1,i,i)x(i) + 2\sum_{j=0}^{i-1} x'(i)R(i+1,i,j)x(j) \\
&+ \sum_{j=0}^{i-1}\sum_{k=0}^{i-1} x'(j)\Delta R(i,j,k)x(k)\Bigg]. \qquad (6.134)
\end{aligned}
$$

As a result from (6.117) via (6.123), (6.128), (6.130)–(6.134), we obtain

$$
\begin{aligned}
\inf_{u\in U} \mathbf{E}\Big[& x'(i)\big[a_1'(0)P_0(i+1)a_1(0) + a_1'(0)Q_0(i+1,i) + Q_0'(i+1,i)a_1(0) \\
&+ R(i+1,i,i) + F_1(i) - P_0(i)\big]x(i) \\
&+ 2x'(i)\big[a_1'(0)P_0(i+1)a_2(i) + a_1'(0)P_1(i+1)
\end{aligned}
$$

$$
\begin{aligned}
&+ Q_0'(i+1,i)a_2(i) + Q_1'(i+1,i) - P_1(i)\big]m(i) \\
&+ m'(i)\big[a_2'(i)P_0(i+1)a_2(i) + a_2'(i)P_1(i+1) + P_1'(i+1)a_2(i) \\
&+ \Delta P_2(i)\big]m(i) + Tr[a_2'(i)P_0(i+1)a_2(i)\gamma(i) + b'(i)P_0(i+1)b(i)] \\
&+ 2Tr[a_2'(i)P_1(i+1)W(i)a_2(i)\gamma(i) + b'(i)P_1(i+1)W(i)b(i)] \\
&+ Tr[a_2'(i)W'(i)P_2(i+1)W(i)a_2(i)\gamma(i) \\
&+ b'(i)W'(i)P_2(i+1)W(i)b(i)\big] + \Delta P_3(i) \\
&+ 2\sum_{j=0}^{i-1} x'(i)\big[a_1'(0)P_0(i+1)a_1(i-j) + R(i+1,i,j) \\
&+ a_1'(0)Q_0(i+1,j) + Q_0'(i+1,i)a_1(i-j) - Q_0(i,j)\big]x(j) \\
&+ 2\sum_{k=0}^{i-1} m'(i)\big[a_2'(i)P_0(i+1)a_1(i-k) + P_1'(i+1)a_1(i-k) \\
&+ a_2'(i)Q_0(i+1,k) + \Delta Q_1(i,k)\big]x(k) \\
&+ \sum_{j=0}^{i-1}\sum_{k=0}^{i-1} x'(j)\big[a_1'(i-j)P_0(i+1)a_1(i-k) + a_1'(i-j)Q_0(i+1,k) \\
&+ Q_0'(i+1,j)a_1(i-k) + \Delta R(i,j,k)\big]x(k) + \mu(u(i))\Bigg] = 0, \qquad (6.135)
\end{aligned}
$$

where

$$\mu(u(i)) = u'(i)Z(i)u(i) + 2u'(i)a_0'(i)S(m(i),x_i), \qquad (6.136)$$

and

$$
\begin{aligned}
S(m(i),x_i) &= Z_0(i)m(i) + \sum_{j=0}^{i} Z_1(i,j)x(j), \\
Z(i) &= G(i) + a_0'(i)P_0(i+1)a_0(i), \\
Z_0(i) &= P_0(i+1)a_2(i) + P_1(i+1), \\
Z_1(i) &= P_0(i+1)a_1(i-j) + Q_0(i+1,j). \qquad (6.137)
\end{aligned}
$$

Calculating the infimum of $\mu(u(i))$, from (6.136) we obtain the optimal control of the control problem (6.112), (6.113) in the final form

$$u_0(i) = -Z^+(i)a_0'(i)S(m(i),x_i), \quad i = 0,1,\ldots,N-1. \qquad (6.138)$$

6.4.3 The Optimal Cost Construction

Put $\hat{a}_0 = a_0(i)Z^+(i)a_0'(i)$. Substituting (6.138) into (6.136), we have

$$\begin{aligned}
\mu_0(u_0(i)) &= S'(m(i), x_i)\hat{a}_0(i)S(m(i), x_i) - 2S'(m(i), x_i)\hat{a}_0(i)S(m(i), x_i) \\
&= -S'(m(i), x_i)\hat{a}_0(i)S(m(i), x_i) \\
&= -\left(Z_0(i)m(i) + Z_1(i, i)x(i) + \sum_{j=0}^{i-1} Z_1(i, j)x(j)\right)' \hat{a}_0(i) \\
&\quad \times \left(Z_0(i)m(i) + Z_1(i, i)x(i) + \sum_{k=0}^{i-1} Z_1(i, k)x(k)\right) \\
&= -x'(i)Z_1'(i, i)\hat{a}_0(i)Z_1(i, i)x(i) - 2m'(i)Z_0'(i)\hat{a}_0(i)Z_1(i, i)x(i) \\
&\quad - m'(i)Z_0'(i)\hat{a}_0(i)Z_0(i)m(i) - 2\sum_{k=0}^{i-1} x'(i)Z_1'(i, i)\hat{a}_0(i)Z_1(i, k)x(k) \\
&\quad - 2\sum_{k=0}^{i-1} m'(i)Z_0'(i)\hat{a}_0(i)Z_1(i, k)x(k) \\
&\quad - \sum_{j=0}^{i-1}\sum_{k=0}^{i-1} x'(j)Z_1'(i, j)\hat{a}_0(i)Z_1(i, k)x(k).
\end{aligned} \tag{6.139}$$

Substituting (6.139) into (6.135), we obtain the following recurrence formulas for $P_0(i)$, $P_1(i)$, $P_2(i)$, $P_3(i)$, $Q_0(i, j)$, $Q_1(i, j)$ and $R(i, j, k)$:

$$\begin{aligned}
P_0(i) &= a_1'(0)P_0(i+1)a_1(0) + a_1'(0)Q_0(i+1, i) + Q_0'(i+1, i)a_1(0) \\
&\quad + R(i+1, i, i) + F_1(i) - Z_1'(i, i)\hat{a}_0(i)Z_1(i, i), \\
P_1(i) &= a_1'(0)P_0(i+1)a_2(i) + a_1'(0)P_1(i+1) \\
&\quad + Q_0'(i+1, i)a_2(i) + Q_1'(i+1, i) - Z_1'(i, i)\hat{a}_0(i)Z_0'(i), \\
P_2(i) &= P_2(i+1) + a_2'(i)P_0(i+1)a_2(i) \\
&\quad + a_2'(i)P_1(i+1) + P_1'(i+1)a_2(i) - Z_0'(i)\hat{a}_0(i)Z_0(i), \\
P_3(i) &= P_3(i+1) + Tr[a_2'(i)P_0(i+1)a_2(i)\gamma(i) + b'(i)P_0(i+1)b(i) \\
&\quad + 2a_2'(i)P_1(i+1)W(i)a_2(i)\gamma(i) + 2b'(i)P_1(i+1)W(i)b(i) \\
&\quad + a_2'(i)W'(i)P_2(i+1)W(i)a_2(i)\gamma(i) + b'(i)W'(i)P_2(i+1)W(i)b(i)], \\
Q_0(i, j) &= a_1'(0)P_0(i+1)a_1(i-j) + a_1'(0)Q_0(i+1, j) + Q_0'(i+1, i)a_1(i-j) \\
&\quad + R(i+1, i, j) - Z_1'(i, i)\hat{a}_0(i)Z_1(i, j),
\end{aligned}$$

$$\begin{aligned}
Q_1(i,k) &= Q_1(i+1,k) + a_2'(i)P_0(i+1)a_1(i-k) + a_2'(i)Q_0(i+1,k) \\
&\quad + P_1'(i+1)a_1(i-k) - Z_0'(i)\hat{a}_0(i)Z_1(i,k), \\
R(i,j,k) &= R(i+1,j,k) + a_1'(i-j)P_0(i+1)a_1(i-k) + a_1'(i-j)Q_0(i+1,k) \\
&\quad + Q_0'(i+1,j)a_1(i-k) - Z_1'(i,j)\hat{a}_0(i)Z_1(i,k).
\end{aligned} \tag{6.140}$$

Here

$$i = N-1, N-2, \ldots, 1, 0, \qquad j,k = 0,1,\ldots,i-1.$$

By virtue of (6.122), (6.137) and (6.140) the functions $P_0(i)$, $P_1(i)$, $P_2(i)$, $P_3(i)$, $Q_0(i,j)$, $Q_1(i,j)$ and $R(i,j,k)$ can be calculated for all $i = 0,1,\ldots,N$, $j,k = 0,1,\ldots,i-1$. From this the optimal control and the optimal cost of the control problem (6.112), (6.113) can be obtained by virtue of (6.137), (6.138) and (6.120).

6.4.4 A Particular Case

Consider the control problem (6.112), (6.113) in the particular case $a_2(i) = 0$. It means that the unknown parameter in the system (6.112) is absent. It is easy to see that in this case $P_1(i) = 0$, $P_2(i) = 0$, $Q_1(i,j) = 0$. So, the system (6.140) takes the form

$$\begin{aligned}
P_0(i) &= a_1'(0)P_0(i+1)a_1(0) + a_1'(0)Q_0(i+1,i) + Q_0'(i+1,i)a_1(0) \\
&\quad + R(i+1,i,i) + F_1(i) - Z_1'(i,i)\hat{a}_0(i)Z_1(i,i), \\
P_3(i) &= P_3(i+1) + Tr[b'(i)P_0(i+1)b(i)], \\
Q_0(i,j) &= a_1'(0)P_0(i+1)a_1(i-j) + a_1'(0)Q_0(i+1,j) + Q_0'(i+1,i)a_1(i-j) \\
&\quad + R(i+1,i,j) - Z_1'(i,i)\hat{a}_0(i)Z_1(i,j), \\
R(i,j,k) &= R(i+1,j,k) + a_1'(i-j)P_0(i+1)a_1(i-k) + a_1'(i-j)Q_0(i+1,k) \\
&\quad + Q_0'(i+1,j)a_1(i-k) - Z_1'(i,j)\hat{a}_0(i)Z_1(i,k), \\
&\quad i = N-1, N-2, \ldots, 1, 0, \quad j,k = 0,1,\ldots,i-1,
\end{aligned} \tag{6.141}$$

and the optimal cost $J(u_0)$ via (6.120), (6.121) equals

$$\begin{aligned}
J(u_0) &= \mathbf{E}V(0, x(0)) \\
&= \mathbf{E}[x'(0)P_0(0)x(0)] + P_3(0).
\end{aligned} \tag{6.142}$$

Example 6.3 Consider the scalar optimal control problem for the equation

$$x(i+1) = a_0u(i) + x(i) + a_1x(i-1) + b\xi(i+1), \tag{6.143}$$

and the performance functional

$$J(u) = \mathbf{E}\left[x^2(N) + \lambda \sum_{i=0}^{N-1} u^2(i)\right]. \tag{6.144}$$

It is easy to see that the optimal control problem (6.143), (6.144) is a particular case of the control problem (6.112), (6.113) by

$$\begin{gathered} a_0(0) = a_0, \quad a_0(i) = 0, \quad i > 0, \\ a_1(0) = 1, \quad a_1(1) = a_1, \quad a_1(i) = 0, \quad i > 1, \\ a_2(i) = 0, \quad b(i) = b, \quad i \geq 0, \\ F = 1, \quad F_1(i) = 0, \quad G(i) = \lambda, \quad i \geq 0. \end{gathered}$$

Solving the system (6.141) with the bound conditions

$$P_0(N) = 1, \quad P_3(N) = 0, \quad R(N, j, k) = 0, \quad Q_0(N, j) = 0$$

and the following values of the parameters

$$N = 15, \quad a_0 = 0.01, \quad a_1 = 0.15, \quad b = 0.2, \quad \lambda = 1,$$

for the optimal control problem (6.143), (6.144) we obtain via (6.138) and (6.137) the optimal control

$$\begin{aligned} u_0(0) &= -0.2921x(0), \\ u_0(1) &= -0.2284x(1) - 0.0302x(0), \\ u_0(2) &= -0.1784x(2) - 0.0236x(1), \\ u_0(3) &= -0.1393x(3) - 0.0185x(2), \\ u_0(4) &= -0.1089x(4) - 0.0144x(3), \\ u_0(5) &= -0.0849x(5) - 0.0112x(4), \\ u_0(6) &= -0.0663x(6) - 0.0088x(5), \\ u_0(7) &= -0.0517x(7) - 0.0068x(6), \\ u_0(8) &= -0.0403x(8) - 0.0053x(7), \\ u_0(9) &= -0.0315x(9) - 0.0042x(8), \\ u_0(10) &= -0.0245x(10) - 0.0033(9), \\ u_0(11) &= -0.0191x(11) - 0.0025x(10), \\ u_0(12) &= -0.0149x(12) - 0.0020x(11), \\ u_0(13) &= -0.0115x(13) - 0.0015x(12), \\ u_0(14) &= -0.0100x(14) - 0.0015x(13), \end{aligned}$$

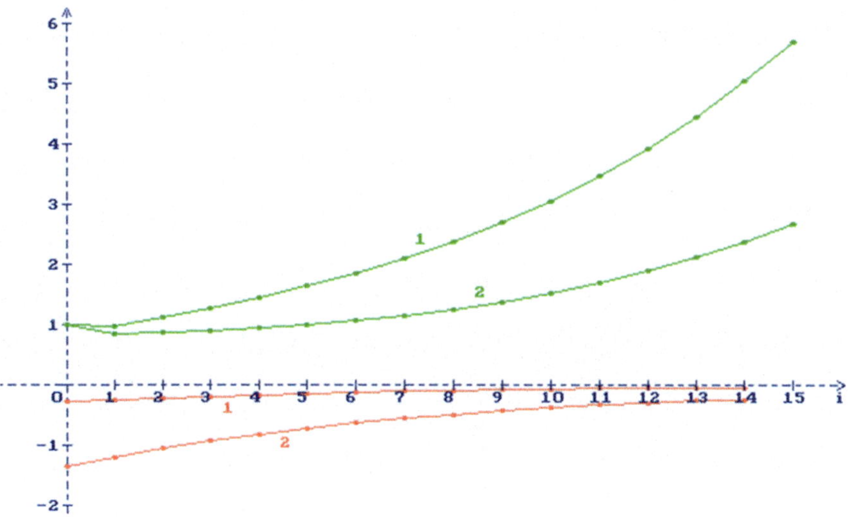

Fig. 6.1 Two trajectories of the optimal solution (*green*) and the optimal control (*red*) for the values of the parameters $N = 15, a_1 = 0.15, \lambda = 1, b = 0$, (*1*) $a_0 = 0.01$, $J(u_0) = 33.0793$, (*2*) $a_0 = 0.1$, $J(u_0) = 15.5058$

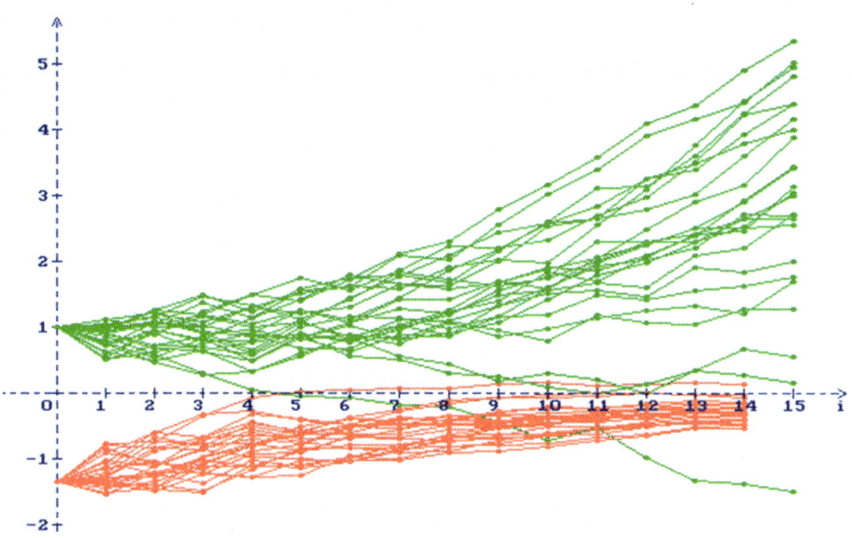

Fig. 6.2 25 trajectories of the optimal solution (*green*) and the optimal control (*red*) for the values of the parameters $N = 15, a_0 = 0.1, a_1 = 0.15, \lambda = 1, b = 0.2$

and via (6.142) the optimal cost

$$J(u_0) = 33.0793\mathbf{E}x^2(0) + 4.6090.$$

In Fig. 6.1 two trajectories of the optimal solution $x_0(i)$ (green) and the optimal control $u_0(i)$ (red) are shown for the initial condition $x(0) = 1$ and the following values of the parameters $N = 15$, $a_1 = 0.15$, $\lambda = 1$, $b = 0$ (deterministic case): (1) $a_0 = 0.01$, $J(u_0) = 33.0793$; (2) $a_0 = 0.1$, $J(u_0) = 15.5058$.

In Fig. 6.2 25 trajectories of the optimal solution $x_0(i)$ (green) and the optimal control $u_0(i)$ (red) are shown for the initial condition $x(0) = 1$ and the following values of the parameters $N = 15$, $a_0 = 0.1$, $a_1 = 0.15$, $\lambda = 1$, $b = 0.2$ (stochastic case). For simulation of stochastic perturbations $\xi(i)$ in the Eq. (6.143), a sequence of random variables uniformly distributed on $(-\sqrt{3}, \sqrt{3})$ was used (at that $\mathbf{E}\xi(i) = 0$, $\mathbf{E}\xi^2(i) = 1$). Calculations are shown that 25 different values of the functional

$$J(u_0, \xi) = x_0^2(N) + \sum_{i=0}^{N-1} u_0^2(i)$$

belong to the interval (4.4722, 44.1813). At that the optimal cost obtained by virtue of (6.141), (6.142) (and the given initial condition $x(0) = 1$) equals

$$\begin{aligned} J(u_0) &= \mathbf{E}J(u_0, \xi) \\ &= 15.5058x^2(0) + 3.2035 \\ &= 18.7093. \end{aligned}$$

References

1. Aceto L, Pandolfi R, Trigiante D (2006) One parameter family of linear difference equations and the stability problem for the numerical solution of ODEs. Adv Differ Equ 2006:1–14
2. Agarwal RP (2000) Difference equations and inequalities. Theory, methods and applications. Dekker, New York
3. Agarwal RP, Pituk M (2007) Asymptotic expansions for higher-order scalar difference equations. Adv Differ Equ 2007:12. doi:10.1155/2007/67492
4. Ahmed HM (2008) On some fractional stochastic integral equations. Int J Math Anal 2(7):299–306
5. Alafif HA (2014) Random integral equation of the Volterra type with applications. J Appl Math Phys 2:138–149. http://dx.doi.org/10.4236/jamp.2014.25018
6. Allen LJS (1994) Some discrete-time SI, SIR, and SIS epidemic models. Math Biosci 124:83–105
7. Allen LJS, Jones MA, Martin CF (1991) A discrete-time model with vaccination for a measles epidemic. Math Biosci 105:111–131
8. d'Andrea-Novel B, De Lara M (2013) Control theory for engineers, vol XV. Springer, London
9. Andreeva EA, Kolmanovskii VB, Shaikhet LE (1992) Control of hereditary systems. Nauka, Moscow, 336 pp (in Russian)
10. Appleby JAD, Gyori I, Reynolds DW (2006) On exact convergence rates for solutions of linear equations of Volterra difference equations. J Differ Equ Appl 12:1257–1275
11. Arino O, Elabdllaoui A, Mikaram J, Chattopadhyay J (2004) Infection on prey population may act as a biological control in a ratio-dependent predator-prey model. Nonlinearity 17:1101–1116
12. Äström KJ, Wittenmark B (1984) Computer-controlled systems. Prentice-Hall, New York
13. Athans M, Falb PL (1966) Optimal control: an introduction to the theory and applications. McGraw-Hill, New York
14. Aoki M (1967) Optimization of stochastic systems. Academic Press, New York
15. Bailey NTJ (1957) The mathematical theory of epidemics. Griffin, London
16. Bao J, Lee PL (2007) Process control. The passive systems approach. Advances in industrial control. Springer, London
17. Bashkov AB, Kolmanovskii VB, Mao X, Matasov AI (2004) Mean-square filtering problem for discrete Volterra equations. Stochast Anal Appl 22(4):1085–1110
18. Bashkov A, De Nicolao G, Kolmanovskii V, Matasov A (2007) Filtering problem for discrete Volterra equations with combined disturbances. Stochast Anal Appl 25(6):1297–1323
19. Bellman R, Cooke CL (1963) Differential-difference equations. Mathematics in science and engineering. The Rand Corporation, Santa Monica

L. Shaikhet, *Optimal Control of Stochastic Difference Volterra Equations*,
Studies in Systems, Decision and Control 17, DOI 10.1007/978-3-319-13239-6

20. Bensoussan A, Runggaldier W (1987) An approximation method for stochastic control problems with partial observation of the state—a method for constructing ε-optimal controls. Actu Applicandae Mathematicae 10:145–170
21. Beretta E, Takeuchi Y (1995) Global stability of an SIR epidemic model with time delays. J Math Biol 33:250–260
22. Beretta E, Kolmanovskii V, Shaikhet L (1998) Stability of epidemic model with time delays influenced by stochastic perturbations. Math Comput Simul 45(3–4):269–277 (Special Issue "Delay Systems")
23. Biran Y, Innis B (1979) Optimal control of bilinear systems: time-varying effects of cancer drugs. Automatica 15:325–329
24. Blair WP, Sworder DD (1975) Feedback control of a class of linear discrete systems with jump parameters and quadratic cost criteria. Int J Control 21(5):833–841
25. Blizorukov MG (1996) On the construction of solutions of linear difference systems with continuous time. Differ Uravn (Minsk) 32:127–128. Transl Differ Equ 133–134
26. Borne P, Kolmanovskii V, Shaikhet L (1999) Steady-state solutions of nonlinear model of inverted pendulum. Theory Stochast Process 5(21)(3–4):203–209. In: Proceedings of the third Ukrainian-Scandinavian conference in probability theory and mathematical statistics, Kyiv, Ukraine, 8–12 June 1999
27. Borne P, Kolmanovskii V, Shaikhet L (2000) Stabilization of inverted pendulum by control with delay. Dyn Syst Appl 9(4):501–514
28. Boltyanskii VG (1966) Mathematical methods of optimal control. Nauka, Moscow (in Russian)
29. Boltyanskii VG (1973) The optimal control of discrete systems. Nauka, Moscow (in Russian)
30. Botan C, Ostafi F (2008) A solution to the continuous and discrete-time linear quadratic optimal problems. WSEAS Trans Syst Control 3(2):71–78
31. Bradul N (2013) Optimal estimation problem. Proc IPMM NANU 26:21–30 (in Russian)
32. Bradul N, Shaikhet L (2004) Optimal stabilization problem for stochastic difference Volterra equation. Proc IPMM NANU 9:24–45 (in Russian)
33. Bradul N, Shaikhet L (2007) Stability of the positive point of equilibrium of Nicholson's blowflies equation with stochastic perturbations: numerical analysis. Discrete Dyn Nat Soc 2007(92959):25. doi:10.1155/2007/92959
34. Bradul N, Shaikhet L (2009) Stability of difference analogue of mathematical predator-prey model by stochastic perturbations. Vest Odesskogo Naz Univ Mat Mekh 14(20):7–23 (in Russian)
35. Brunner H (2004) The numerical analysis of functional integral and integro-differential equations of Volterra type. Acta Numerica 13:55–145.
36. Brunner H (2009) Recent advances in the numerical analysis of Volterra functional differential equations with variable delays. J Comput Appl Math 228(2):524–537
37. Brunner H, Van der Houwen PJ (1986) The numerical solution of Volterra equations. CWI monographs, vol 3. North-Holland, Amsterdam
38. Buhmann M, Iserles A (1992) Numerical analysis of functional equations with a variable delay. Numerical analysis. Longman, Harlow
39. Buhmann M, Iserles A (1992) On the dynamics of a discretized neutral equation. IMA J Numer Anal 12:339–363
40. Bukhgeim AL (1983) Volterra equations and inverse problems. Nauka, Novosibirsk (in Russian)
41. Burton TA (1983) Volterra integral and difference equations. Academic Press, New York
42. Caraballo T, Shaikhet L (2014) Stability of delay evolution equations with stochastic perturbations. Commun Pure Appl Anal 13(5):2095–2113. doi:10.3934/cpaa.2014.13.2095
43. Caravani P (2013) Modern linear control design. A time-domain approach. Springer, Berlin
44. Castillo-Chavez C, Yakubu AA (2001) Discrete-time S-I-S models with complex dynamics. Nonlinear Anal 47:4753–4762
45. Chang JL (2008) Combining state estimator and disturbance observer in discrete time sliding mode controller design. Asian J Control 10(5):515–524

46. Che WW, Yang GH (2008) State feedback H1 control for quantized discrete-time systems. Asian J Control 10(6):718–723
47. Chernous'ko FL, Kolmanovskii VB (1978) The optimal control under random perturbations. Nauka, Moscow (in Russian)
48. Chernous'ko FL, Kolmanovskii VB (1979) Computational and approximate methods of optimal control. J Soviet Math 12:310–353
49. Crisci MR, Kolmanovskii VB, Russo E, Vecchio A (1995) Stability of continuous and discrete Volterra integro-differential equations by Lyapunov approach. J Integr Equ 4(7):393–411
50. Crisci MR, Kolmanovskii VB, Russo E, Vecchio A (1998) Stability of difference Volterra equations: direct Lyapunov method and numerical procedure. Comput Math Appl 36(10–12):77–97
51. Crisci MR, Kolmanovskii VB, Russo E, Vecchio A (2000) On the exponential stability of discrete Volterra equations. J Differ Equ Appl 6:667–680
52. Cuevas C, Dantas F, Choquehuanca M, Soto H (2013) l^p-boundedness properties for Volterra difference equations. Appl Math Comput 219:6986–6999
53. Cushing JM (1977) Integro-differential equations and delay models in population dynamics. Lectures notes in biomathematics. Springer, Berlin
54. Desch W, Londen SO (2011) An L_p-theory for stohastic integral equations. J Evol Equ 11(2):287–317
55. Elaydi SN (1994) Periodicity and stability of linear Volterra difference systems. J Differ Equ Appl 181(2):483–492
56. Elaydi SN (2005) An introduction to difference equations, 3rd edn. Springer, Berlin
57. Elaydi SN (2009) Stability and asymptoticity of Volterra difference equations: a progress report. J Comput Appl Math 228:504–513
58. Elaydi SN, Murakami S (1996) Asymptotic stability versus exponential stability in linear Volterra difference equations of convolution type. J Differ Equ Appl 2:401–410
59. Elaydi SN, Cushing J, Lasser R, Papageorgiou V, Ruffing A (2007) Difference equations, special functions and orthogonal polynomials. World Scientific, Hackensack
60. El-Borai MM, El-Said El-Nadi Kh, Mostafa OL, Ahmed HM (2004) Volterra equations with fractional stochastic integrals. Math Probl Eng 2004(5):453–468. http://dx.doi.org/10.1155/S1024123X04312020
61. El-Dessoky MM, Yassen MT, Aly ES (2014) Bifurcation analysis and chaos control in Shimizu-Morioka chaotic system with delayed feedback. Appl Math Comput 243:283–297
62. Enatsu Y, Nakata Y, Muroya Y (2010) Global stability for a class of discrete SIR epidemic models. Math Biol Eng 2:349–363
63. Fleming W, Rishel R (1975) Deterministic and stochastic optimal control. Springer, New York
64. Ford NJ, Edwards JT, Roberts JA, Shaikhet LE (1997) Stability of a difference analogue for a nonlinear integro differential equation of convolution type. Numerical analysis report 312, University of Manchester
65. Fridman E (2001) New Lyapunov-Krasovskii functionals for stability of linear retarded and neutral type systems. Syst Control Lett 43:309–319
66. Fridman E (2002) Stability of linear descriptor systems with delays: a Lyapunov-based approach. J Math Anal Appl 273:24–44
67. Gikhman II, Skorokhod AV (1969) Introduction to the theory of random processes. Saunders, Philadelphia
68. Gikhman II, Skorokhod AV (1969) Controlled stochastic processes. Saunders, Philadelphia
69. Gikhman II, Skorokhod AV (1972) Stochastic differential equations. Springer, Berlin
70. Gikhman II, Skorokhod AV (1974) The theory of stochastic processes, vol I. Springer, Berlin
71. Gikhman II, Skorokhod AV (1975) The theory of stochastic processes, vol II. Springer, Berlin
72. Gikhman II, Skorokhod AV (1979) The theory of stochastic processes, vol III. Springer, Berlin
73. Golec J, Sathananthan S (2001) Strong approximations of stochastic integro-differential equations. Dyn Continuous Discrete Impuls Syst Ser B Appl Algorithms 8(1): 139–151
74. Gonzalez-Sanchez D, Hernandez-Lerma O (2013) Discretetime stochastic control and dynamic potential games. The Euler-equation approach. Springer briefs in mathematics. Springer, New York

75. Gopalsamy K (1992) Equations of mathematical ecology. Kluwer, Dordrecht
76. Gyori I, Reynolds DW (2010) On admissibility of the resolvent of discrete Volterra equations. J Differ Equ Appl 16:1393–1412
77. Halanay A, Morozan T (1992) Stabilization by linear feedback of linear discrete stochastic systems. Rev Roum Math Pures et Appl 37:213–224
78. Hansson A (2000) A primal-dual interior-point method for robust optimal control of linear discrete-time systems. IEEE Trans Autom Control AC-45(8):1532–1536
79. Hasegawa Y (2013) Control problems of discrete-time dynamical systems. Lecture notes in control and information sciences, vol XII. Springer, Berlin, p 447
80. Hocking L (1991) Optimal control: an introduction to the theory with applications. Oxford University Press, Oxford
81. Hromadka II TV, Whitley RJ (1989) Stochastic integral equations and rainfall-runoff models. Springer, Berlin. ISBN: 978-3-642-49311-9 (Print) 978-3-642-49309-6 (Online)
82. Huang Y, Zang W, Zang H (2008) Infinite horizon linear quadratic optimal control for discrete-time stochastic systems. Asian J Control 10(5):608–615
83. Iserles A (1994) Numerical analysis of delay differential equations with variable delays. Ann Numer Math 1:133–152
84. Isidori A (1995) Nonlinear control systems. Communications and control engineering. Springer, London
85. Jager B, Keulen T, Kessels J (2013) Optimal control of hybrid vehicles. Advances in industrial control, vol XVIII. Springer, London
86. Jiang Y, Wei H, Song X, Mei L, Su G, Qui Sh (2009) Global attractivity and permanence of a delayed SVEIR epidemic model with pulse vaccination and saturation incidence. Appl Math Comput 213:312–321
87. Joshi HR (2002) Optimal control of an HIV immunology model. Opt Control Appl Methods 23:199–213
88. Joshi HR, Lenhart S, Gaff H (2006) Optimal harvesting in an integro-difference population model. Opt Control Appl Methods 27:61–75
89. Joshi HR, Lenhart S, Lou H, Gaff H (2007) Harvesting control in an integro-difference population model with concave growth term. Nonlinear Anal Hybrid Syst 1:417–429
90. Karafyllis I, Jiang Z-P (2011) Stability and stabilization of nonlinear systems. Communications and control engineering. Springer, Berlin
91. Karafyllis I, Kotsios S (2006) Necessary and sufficient conditions for robust global asymptotic stabilization of discrete-time systems. J Differ Equ Appl 12(7):741–768
92. Karafyllis I, Kravaris C (2007) On the observer problem for discrete-time control systems. IEEE Trans Autom Control 52(1):12–25
93. Karczewska A (2003) On the limit measure to stochastic Volterra equations. J Integr Equ Appl 15(1):59–77
94. Karczewska A, Zabczyk J (2000) Regularity of solutions to stochastic Volterra equations. Rend Mat Accad Lincei 11:141–154
95. Kelley W, Peterson AC (2001) Difference equations: an introduction with applications. Academic Press, New York
96. Kharitonov V (2013) Time-delay systems. Lyapunov functionals and matrices. Control engineering. Springer, Berlin
97. Kleptsina ML, Veretennikov AYu (1985) On filtering and properties of Ito-Volterra process. Statistics and control of stochastic process: optimization software. IBM Inc., Publication Division, New York, pp 179–196
98. Kocic VL, Ladas G (1993) Global behavior of nonlinear difference equations of higher order with applications. Kluwer, Dordrecht
99. Kolmanovskii VB (1997) Problems of optimal control. Soros Educ J 6:121–127 (in Russian)
100. Kolmanovskii VB (1999) Problems of optimal estimation. Soros Educ J 11:122–127 (in Russian)
101. Kolmanovskii VB, Castellanos-Velasco E, Torres-Munoz JA (2003) A survey: stability and boundedness of Volterra difference equations. Nonlinear Anal Theory Methods Appl 53:861–928

102. Kolmanovskii VB, Kosareva NP (2001) Stability of Volterra difference equations. Differ Equ 37(12):1773–1782
103. Kolmanovskii VB, Myshkis AD (1992) Applied theory of functional differential equations. Kluwer, Dordrecht
104. Kolmanovskii VB, Myshkis AD (1999) Introduction to the theory and applications of functional differential equations. Kluwer, Dordrecht
105. Kolmanovskii VB, Nosov VR (1981) Stability and periodical regimes of regulating hereditary systems. Nauka, Moscow (in Russian)
106. Kolmanovskii VB, Nosov VR (1986) Stability of functional differential equations. Academic Press, London
107. Kolmanovskii VB, Shaikhet LE (1989) States estimate of hereditary stochastic systems. Stochast Anal Appl 7(4):387–411
108. Kolmanovskii VB, Shaikhet LE (1995) General method of Lyapunov functionals construction for stability investigations of stochastic difference equations. Dynamical systems and applications. World Scientific series in applicable analysis, vol 4. World Scientific, New Jersey, pp 397–439
109. Kolmanovskii VB, Shaikhet LE (1996) Control of systems with aftereffect. Translations of mathematical monographs, vol. 157. American Mathematical Society, Providence
110. Kolmogorov AN, Fomin SV (1957) Elements of the theory of functions and functional analysis. Graylock Press, New York
111. Kolokol'tsov VN, Schilling RL, Tyukov AE (2004) Estimates for multiple stochastic integrals and stochastic Hamilton-Jacobi equations. Rev Mat Iberoamericana 20:333–380
112. Krasovskii NN (1968) The theory of motion control. Nauka, Moscow
113. Krasovskii NN (1985) Dynamic system control. Nauka, Moscow
114. Krylov NV (1980) Controlled diffusion processes. Springer, New York
115. Kuang Y (1993) Delay differential equations with application in population dynamics. Academic Press, New York
116. Kuchkina N (1998) Optimal control with incomplete data. Theory Stochast Process 4(20)1–2:161–171
117. Kuchkina N, Shaikhet L (1996) Stability of the stochastic difference Volterra equations. Theory Stochast Process 2(18)3–4:79–86
118. Kuchkina N, Shaikhet L (1997) Optimal control problem for nonlinear stochastic difference second kind Volterra equations. Comput Math Appl 34(9):65–73
119. Kuchkina N, Shaikhet L (1998) Optimal control of Volterra type stochastic difference equations. Comput Math Appl 36(10–12):251–259 (Special Issue "Advances in difference equations II")
120. Kuchkina N, Shaikhet L (1998) Optimal estimation of stochastic difference equations. In: Proceeding of CESA'98. Symposium on signal processing and cybernetics, vol 4. Tunisia, April 1–4, pp 165–169
121. Kulenovic MRS, Merino O (2002) Discrete dynamical systems and difference equations with mathematica. Chapman and Hall/CRC, Boca Raton
122. Kushner HJ (1967) Stochastic stability and control. Academic Press, New York
123. Lakshmikantham V, Trigiante D (2002) Theory of difference equations: numerical methods and applications. Dekker, New York
124. Lantos B, Merton L (2011) Nonlinear control of vehicles and robots. Advances in Industrial Control. Springer, London
125. Lavretsky E, Wise K (2013) Robust and adaptive control with aerospace applications. Advanced textbooks in control and signal processing. Springer, London
126. Levin JJ, Nohel JA (1968) The integrodifferential equations of a class of nuclear reactors with delayed neutrons 31:151–172
127. Li J, Ma Z, Brauer F (2007) Global analysis of discrete-time SI and SIS epidemic models. Math Biosci Eng 4:699–710
128. Liao LZ, Li D (2000) Successive method for general multiple linear-quadratic control problem in discrete time. IEEE Trans Autom Control AC-45(7):1380–1385

129. Linz P (1985) Analytical and numerical methods for Volterra equations. SIAM, Philadelphia
130. Liptser RSh, Shiryayev AN (1977) Statistics of random processes. Springer, New York
131. Liu B, Su H, Li R, Sun D, Hu W (2014) Switching controllability of discrete-time multi-agent systems with multiple leaders and time-delays. Appl Math Comput 228:571–588
132. Macdonald N (1978) Time lags in biological models. Lecture notes in biomathematics, vol 27. Springer, Berlin
133. Mackevicius V (2011) Introduction to stochastic analysis: integrals and differential equations. Wiley, Hoboken. ISBN: 978-1-84821-311-1
134. Mackey MC, Glass L (1997) Oscillation and chaos in physiological control system. Science 197:287–289
135. Maistrenko Y, Sharkovsky AN (1986) Difference equations with continuous time as mathematical models of the structure emergences. Dynamical systems and environmental models. Mathematical ecology. Akademie-Verlag, Berlin, pp 40–49
136. Maleknejad K, Khodabin M, Rostami M (2012) A numerical method for solving m-dimensional stochastic Ito-Volterra integral equations by stochastic operational matrix. Comput Math Appl 63(1):133–143
137. Maleknejad K, Khodabin M, Shekarabi FH (2014) Modified block pulse functions for numerical solution of stochastic Volterra integral equations. J Appl Math 2014(Article ID 469308):10. http://dx.doi.org/10.1155/2014/469308
138. Malek Zavarei M, Jamshide M (1987) Time-delay systems: analysis, optimizations and applications. North Holland, Amsterdam
139. Marotto F (1982) The dynamics of a discrete population model with threshold. Math Biosci 58:123–128
140. Marshall JE (1978) Control of time-delay systems. IEEE control engineering series, vol 10
141. Marshall JE, Salehi SV (1982) Improvement of system performance by the use of time-delay elements. Proc IEEE-D 129(5):177
142. Medhin NG (1986) Optimal processes governed by integral equations. J Math Anal Appl 120(1):1–12
143. Medina R (2001) Asymptotic behavior of Volterra difference equations. Comput Math Appl 41(5–6):679–687
144. Menshutkin VV (1971) Mathematical modelling of populations and associations of water animals. Nauka, Leningrad (in Russian)
145. Mikosch T, Norvaisa R (2000) Stochastic integral equations without probability. Ber-noulli 6(3):401–434
146. Milman MM (1987) Representation for the optimal control law for stable hereditary systems. Distributed parameter systems. Lecture notes in control and information science, vol 102. Springer, Heidelberg
147. Mokkedem FZ, Fu X (2014) Approximate controllability of semi-linear neutral integro-differential systems with finite delay. Appl Math Comput 242:202–215
148. Mollison D (2003) Epidemic models: their structure and relation to data. Publications of the Newton Institute, Cambridge University Press, Cambridge
149. Morozan T (1983) Stabilization of some stochastic discrete-time control systems. Stochast Anal Appl 1(1):89–116
150. Morozan T (1979) Stochastic stability and control for discrete-time systems with jump Markov disturbances. Rev Roum Math Pures et Appl 24(1):111–127
151. Muroya Y, Ishiwata E (2009) Stability for a class of difference equations. J Comput Appl Math 228:561–570
152. Murray JD (2002) Mathematical biology, 3rd edn. Springer, New York
153. Myshkis AD (1951) General theory of differential equations with delay. Translations of mathematical monographs, vol 55. American Mathematical Society, Providence
154. Myshkis AD (1972) Linear differential equations with delay argument. Nauka, Moscow (in Russian)
155. Naidu DS, Rao AK (1985) Singular perturbation analysis of discrete control systems. Springer, Berlin

156. Ngoc PHA, Hieu L (2013) New criteria for exponential stability of nonlinear difference systems with time-varying delay. Int J Control 86(9):1646–1651 doi:10.1080/00207179.2013.792004
157. Ngoc PHA, Naito T, Shin JS, Murakami S (2008) Stability and robust stability of positive linear Volterra difference equations. Int J Robust Nonlinear Control 19:552–568
158. Oguztörelli MN (1996) Time-lag control systems. Academic Press, New York
159. Okonkwo ZC (1996) Admissibility and optimal control for difference equations. Dyn Syst Appl 5 (1996) 4:627–634
160. Okonkwo ZC (2000) Admissibility and optimal control for stochastic difference equations. Integral methods in science and engineering. Chapman & Hall/CRC, Houghton, 1998, pp 263–267. Research notes in mathematics, vol 418, Chapman & Hall/CRC, Boca Raton, 2000
161. Okuyama Y (2014) Discrete control systems. Springer, London
162. Pang G, Chen L (2007) A delayed SIRS epidemic model with pulse vaccination. Chaos Solitons Fractals 34:1627–1635
163. Pathak S, Maiti A (2012) Pest control using virus as control agent: A mathematical model. Nonlinear Anal Model Control 17(1):67–90
164. Petersen IR, Matthew RJ, Dupuis P (2000) Minimax optimal control of stochastic uncertain systems with relative entropy constraints. IEEE Trans Autom Control 45:398–412
165. Pontryagin LS, Boltyanskii VG, Gamkrelidze RV, Mishchenko EF (1986) Mathematical theory of optimal processes. Gordon and Breach, New York
166. Roach GF (ed) (1984) Mathematics in medicine and biomechanics. Shiva, Nautwick
167. Sekiguchi M (2009) Permanence of some discrete epidemic models, Int J Biomath 2:443–461
168. Santonja F-J, Shaikhet L (2012) Analysing social epidemics by delayed stochastic models. Discrete Dyn Nat Soc 2012(530472):13. doi:10.1155/2012/530472
169. Santonja F-J, Shaikhet L (2014) Probabilistic stability analysis of social obesity epidemic by a delayed stochastic model. Nonlinear Anal Real World Appl 17:114–125 http://www.sciencedirect.com/science/journal/aip/14681218
170. Sasagava T (1976) Optimal control of stochastic systems depended on the past. Trans Soc Instrum Control Eng12(3):346–370
171. Shaikhet LE (1984) On the optimal control of Volterra's equations. Probl Control Inf Theory 13(3):141–152
172. Shaikhet LE (1985) On the optimal control of integral-functional equations. Prikladnaya matematika mekhanika 49(6):923–934 (in Russian). Transl J App Math Mech (1985) 49(6):704–712. doi:10.1016/0021-8928(85)90006-1
173. Shaikhet LE (1985) On a necessary condition for optimality of control of stochastic systems. Teorija Veroiyatnostej i Matematicheskaja Statistika 33:104–113 (in Russian). English translation in Theory Prob Math Stat (1986) 33:117–126
174. Shaikhet LE (1986) On the ε-optimal control of quasilinear integral equations. Teoriya Slutchajnikh Protsessov 14:96–101 (in Russian). Transl J Soviet Math (1991) 53(1):101–104. doi:10.1007/BF01104059
175. Shaikhet LE (1986) Optimal control of stochastic integral equations. Lect Notes Control Inf Sci 81:178–187. doi:10.1007/BFb0007095
176. Shaikhet L (1992) Construction of successive approximations to optimal control of stochastic quasilinear integral-functional equations. Math Notes 51(2):196–203
177. Shaikhet L (1998) Stability of predator-prey model with aftereffect by stochastic perturbations. Stab Control Theory Appl 1(1):3–13
178. Shaikhet L (2011) Lyapunov functionals and stability of stochastic difference equations. Springer, London
179. Shaikhet L (2013) Lyapunov functionals and stability of stochastic functional differential equations. Springer, Dordrecht
180. Shaikhet L (2014) Stability of a positive equilibrium state for a stochastically perturbed mathematical model of glassy-winged sharpshooter population. Math Biosci Eng 11(5):1167–1174. doi:10.3934/mbe.2014.11.1167

181. Shaikhet L, Korobeinikov A (2014) Stability of a stochastic model for HIV-1 dynamics within a host. CRM (Centre de Recerca Matematica) Preprint Series Number 1191, 13 pp. http://www.crm.cat/en/Publications/Publications/2014/Pr1191.pdf
182. Shaikhet L, Shafir M (1989) Linear filtering solutions of stochastic integral equations in non-gaussian case. Probl Control Inf Theory 18(6):421–434
183. Sharkovsky AN, Maistrenko YuL, Romanenko EYu (1986) Difference equations and their applications. Mathematics and its applications, vol 250. Kluwer, Dordrecht
184. Shiryaev AN (1996) Probability. Springer, Berlin
185. Shtengol'd ESh, Godin EA, Kolmanovskii VB (1990) The control of the stressed-strained state of vessels and hypertensive disease. Nauka, Moscow (in Russian)
186. Siddique N (2014) Intelligent control. A hybrid approach based on fuzzy logic, neural networks and genetic algorithms. Studies in computational intelligence, vol 517. Springer
187. Sipahi R, Vyhlidal T, Niculescu S-I, Pepe P (2012) Time delay systems: methods, applications and new trends. Lecture notes in control and information sciences, vol 423. Springer
188. Song X, Jiang Y, Wei H (2009) Analysis of a saturation incidence SVEIRS epidemic model with pulse and two time delay. Appl Math Comput 214:381–390
189. Song Y, Baker CTH (2004) Perturbation of Volterra difference equations. J Differ Equ Appl 10:379–397
190. Song Y, Tian H (2007) Periodic and almost periodic solutions of nonlinear discrete Volterra equations with unbounded delay. J Comput Appl Math 205(2):859–870
191. Sun H, Hou L (2014) Composite anti-disturbance control for a discrete-time time-varying delay system with actuator failures based on a switching method and a disturbance observer. Nonlinear Analy Hybrid Syst 14:126–138
192. Swords C, Appleby JAD (2010) Stochastic delay difference and differential equations: applications to financial markets. LAP Lambert Academic Publishing
193. Tsokos CP, Padgett WJ (1971) Random integral equations with applications to stochastic systems. Lecture notes in mathematics vol 233. Springer, Berlin
194. Vasiljev FP (1980) Numerical methods for solving extreme problems. Nauka, Moscow (in Russian)
195. Verlan' AF, Sizikov VS (1986) Integral equations: methods, algorithms, and programs. Naukova Dumka, Kiev (in Russian)
196. Verriest E, Delmotte F, Egerstedt M (2005) Control of epidemics by vaccination. Proc Am Control Conf 2:985–990
197. Volterra V (1909) Equasioni integro-differeziali dell elasticita nel caso della isotropia. Rend R Accad dei Lincei 18(5):50–55
198. Volterra V (1931) Lesons sur la theorie mathematique de la lutte pour la vie. Gauthier-Villars, Paris
199. Vyhlidal T, Lafay J-F, Sipahi R (2014) Delay systems. From theory to numerics and applications. Advances in delays and dynamics, vol 1. Springer
200. Wang Z, Liu Y, Wei G, Liu X (2010) A note on control of a class of discrete-time stochastic systems with distributed delays and nonlinear disturbances. Automatica 46(3):543–548
201. Warga J (1977) Optimal control of differential and functional equations. Academic Press, London
202. Wiener N (1949) Extrapolation, interpolation and smoothing of stationary time series. Wiley, New York
203. Willems J (1969) The analysis of feedback systems. MIT Press, Cambridge
204. Wonham WM (1963) Stochastic problems in optimal control. IEEE Convention Rec 2:114–124
205. Wonham WM (1968) On the separation theorem of stochastic control. SIAM J Control 6(2):312–326
206. Xiao J, Xiang W (2014) New results on asynchronous control for switched discrete-time linear systems under dwell time constraint. Appl Math Comput 242:601–611
207. Xu S, Lam J, Yang CW (2001) Quadratic stability and stabilization of uncertain linear discrete-time systems with state delay. Syst Control Lett 43(2):77–84

208. Yan L, Liu B (2010) Stabilization with optimal performance for dissipative discrete-time impulsive hybrid systems. Adv Differ Equ 2010(278240):14. doi:10.1155/2010/278240
209. Zabczyk J (1976) Stochastic control of discrete-time systems. Control Theory Topics Funct Anal 3:187–224
210. Zhang Y, Wang M, Xu H, Teo KL (2014) Global stabilization of switched control systems with time delay. Nonlinear Anal Hybrid Syst 14:86–98
211. Zhong Q-C (2006) Robust control of time-delay systems. Springer
212. Zhou B (2014) Truncated predictor feedback for time-delay systems. Springer
213. Zouyousefain M, Leela S (1990) Stability results for difference equations of Volterra type. Appl Math Comput 36(1):51–61

Index

L. Shaikhet, *Optimal Control of Stochastic Difference Volterra Equations*,
Studies in Systems, Decision and Control 17, DOI 10.1007/978-3-319-13239-6

The manufacturer's authorised representative in the EU is Springer Nature Customer Service Centre GmbH, Europaplatz 3, 69115 Heidelberg, Germany. If you have any concerns regarding our products, please contact ProductSafety@springernature.com

Printed and bound by CPI Group (UK) Ltd, Croydon, CR0 4YY

15/07/2026

02167634-0002